Lecture Notes in Chemistry

Edited by G. Berthier, M. J. S. Dewar, H. Fischer, K. Fukui, H. Hartmann, H. H. Jaffé, J. Jortner, W. Kutzelnigg, K. Ruedenberg, E. Scrocco, W. Zeil

1

Georges Henry Wagnière

Introduction to

Elementary Molecular Orbital Theory and to Semiempirical Methods

Springer-Verlag
Berlin · Heidelberg · New York 1976

Author
Georges Henry Wagnière
Physikalisch-Chemisches Institut
der Universität Zürich
Rämistraße 76
CH-8001 Zürich

Library of Congress Cataloging in Publication Data

Wagnière, Georges Henry, 1933-
Introduction to elementary molecular orbital theory
and to semiempirical methods.

(Lecture notes in chemistry ; v. 1)
Bibliography: p.
Includes index.
1. Molecular orbitals. I. Title.
QD461.W33 541'.28 76-40002

ISBN-13: 978-3-540-07865-4 e-ISBN-13: 978-3-642-93050-8
DOI: 10.1007/978-3-642-93050-8

Softcover reprint of the hardcover 1st edition 1976

Introduction

These notes summarize in part lectures held regularly at the University of Zurich and, in the Summer of 1974, at the Seminario Latinoamericano de Quimica Cuantica in Mexico. I am grateful to those who have encouraged me to publish these lectures or have contributed to them by their suggestions. In particular, I wish to thank Professor J. Keller of the Universidad Nacional Autonoma in Mexico, Professor H. Labhart and Professor H. Fischer of the University of Zurich, as well as my former students Dr. J. Kuhn, Dr. W. Hug and Dr. R. Geiger.

The aim of these notes is to provide a summary and concise introduction to elementary molecular orbital theory, with an emphasis on semiempirical methods. Within the last decade the development and refinement of ab initio computations has tended to overshadow the usefulness of semiempirical methods. However, both approaches have their justification. Ab initio methods are designed for accurate predictions, at the expense of greater computational labor. The aim of semiempirical methods mainly lies in a semiquantitative classification of electronic properties and in the search for regularities within given classes of larger molecules.

The reader is supposed to have had some previous basic instruction in quantum mechanics, such as is now offered in many universities to chemists in their third or fourth year of study. The bibliography should encourage the reader to consult other texts, in particular also selected publications in scientific journals.

I wish to express my gratitude to Miss H. Böckli who has competently typed the entire manuscript and to Mr. E. Spalinger for the drawings.

Zurich, May 1976

G. Wagnière

Contents

I. The hierarchy of approximations

The nonrelativistic Hamiltonian for a molecular system composed of many nuclei (indices A,B) and many electrons (indices μ,ν) reads (neglecting magnetic interactions):

$$\mathcal{H} = \sum_{\mu}\left(-\frac{\hbar^2}{2m_e}\nabla_\mu^2\right) + \sum_{A}\left(-\frac{\hbar^2}{2M_A}\nabla_A^2\right)$$

$$-\sum_{\mu}\sum_{A}\frac{Z_A e^2}{r_{\mu A}} + \sum_{\mu>\nu}\sum\frac{e^2}{r_{\mu\nu}} + \sum_{A>B}\sum\frac{Z_A Z_B e^2}{R_{AB}}$$

which may be more concisely written

$$\mathcal{H} \equiv T_e + T_n + V_{en} + V_{ee} + V_{nn} \equiv \mathcal{H}_e + T_n + V_{nn}$$

Here T stands for the kinetic energy operator, V for the potential energy operator, the subscript e means "electronic", the subscript n means "nuclear".

1. The Born-Oppenheimer approximation

We seek to solve the time-independent molecular Schrödinger equation

$$\mathcal{H}\Psi(r,R) = W \cdot \Psi(r,R)$$

Due to the great mass difference between electrons and atomic nuclei it proves possible to a satisfactory degree of approximation [1] to treat the degrees of freedom of the electrons, designated collectively by r, separately from those of the nuclei, designated here by R. In this sense the solution $\Psi(r,R)$ may approximately be written as a product of two functions, of which one depends only on the general nuclear coordinates R:

$$\Psi(r,R) = \psi(r,R) \cdot v(R)$$

By neglecting some terms which in general may be shown to be small [1,2], it is thus possible to separate the Schrödinger equation into:

a) An equation for the motion of the electrons, the nuclei remaining fixed at frozen positions R':

$$\mathcal{H}_e \psi_m(r,R') = E_m(R') \cdot \psi_m(r,R')$$

Here m denotes a particular electronic state. The electronic energy $E_m(R')$ depends parametrically on the frozen positions of the nuclei. Often one holds the nuclei fixed in experimentally known equilibrium positions.

b) An equation for the motion of the nuclei in the field of the electrons in a given electronic state m:

$$\left\{T_n + V_{nn}(R) + E_m(R)\right\} v_{mj}(R) = e_{mj} \cdot v_{mj}(R)$$

The electronic energy as a function of nuclear position $E_m(R)$ acts as a potential on the nuclei. For a diatomic molecule in a bonding electronic state $E_m(R) + V_{nn}(R)$ is generally described by a Morse potential. $v_{mj}(R)$ represents a vibrational wavefunction j in the electronic state m.

2. The solution of the electronic problem

We consider the electronic equation

$$\mathcal{H}_e \psi_m(r) \equiv (T_e + V_{en} + V_{ee}) \psi_m(r) = E_m \cdot \psi(r)$$

We no longer explicitly mention the nuclear coordinates R, once we have stated where they have been fixed. Of course we still have a many-body problem to deal with, and the solution of this problem is in general still very cumbersome. In practice it proves only feasible to obtain approximate solutions, and it is the degree of approximation that is the crucial question. Even in simplest cases exact solutions require a quasi-infinite amount of labor. In this sense we distinguish between:

a) The ab initio procedure. It seeks in principle exact solutions. All quantities appearing in the calculation are computed as exactly as numerically possible. If an ab initio solution is still approximate, which in practice it always is, this comes from the fact that the form of the wavefunction has been restricted to facilitate the computation.

b) The semiempirical procedure. It seeks from the start only approximate solutions. The simplifications may be quite drastic, but must always be physically justifiable. One may in this sense further distinguish between

i) simplification of the electronic Hamiltonian $\mathcal{H}_e$ itself by, for instance, leaving out the electronic repulsion term V_{ee} and replacing T_e+V_{en} by an effective Hamiltonian;

ii) neglect of some intermediate quantities or their empirical calibration on atomic data and on test-molecules.

To study large molecules procedure b) is often the only tractable one. The more limited reliability of b) as compared to a) is sometimes also compensated by an increased insight into the interrelation of basic quantities.

3. The subdivision of electrons into different groups

From the chemical point of view the electrons in a molecule may be subdivided into those which take part in the formation of chemical bonds, and those which are largely unaffected by bond formation. The former are generally called valence electrons, the latter atomic core electrons. If the molecule in its equilibrium conformation (i.e. the equilibrium geometry of the atomic nuclei) possesses certain elements of symmetry, such as for instance a plane of symmetry in which lie all atoms of the molecule, the valence electrons may appropriately be further subdivided into σ and π electrons.

From his experience the chemist knows that this subdivision is also physically meaningful. The presence of such π electrons in a molecule influences decisively its reactivity and its spectroscopic properties.

II. Simple Hückel theory of π electrons [3]

The electronic Hamiltonian

$$\mathcal{H}_e = T_e + V_{en} + V_{ee} \qquad \text{may be written}$$

$$= T_\pi + T_\sigma + V_{\pi n} + V_{\sigma n} + V_{\pi\pi} + V_{\pi\sigma} + V_{\sigma\sigma}$$

From it we split off a π electron Hamiltonian

$$\mathcal{H}^\pi = T_\pi + V_{\pi n} + V_{\pi\sigma} + V_{\pi\pi}$$

The σ electrons and the nuclei are assumed frozen into a molecular core:

$$\mathcal{H}^\pi = T_\pi + V_{\pi\,\mathrm{Core}} + V_{\pi\pi} = \sum_{\mu}^{(\pi)} h_{\mathrm{core}}(\mu) + \sum_{\mu>\nu}^{(\pi)} \frac{e^2}{r_{\mu\nu}}$$

We further average $V_{\pi\pi}$ to obtain an effective π Hamiltonian as sum of pseudo-one-electron parts:

$$\mathcal{H}_\pi^{\mathrm{Hückel}} = \sum_{\mu}^{(\pi)} h_{\mathrm{eff}}(\mu)$$

In this approximation each π electron moves in an <u>average field</u> of the core and the other π electrons.

We now want to solve the one-electron equation

$$h_{eff}\,\varphi = \epsilon \cdot \varphi$$

As we no longer have an explicit Schrödinger equation, the solutions depend strongly on the form which we impose on them.

1. <u>The LCAO-MO formalism</u>
   (and the Ritz variational principle)

We expand our one-electron functions or molecular orbitals (MO) as linear combinations of basis functions or atomic orbitals (AO) χ_p and write

$$\varphi = \sum_{p=1}^{N} c_p \chi_p$$

In our present case the χ_p are $2p_\pi$ ($\equiv 2p_z$) atomic orbitals centered on each atom contributing one (or possibly two) electron(s) to the π system. The total number of such atoms is N and the index p also denotes a given atom. The expectation value for the one-electron energy ϵ then takes the form

$$\epsilon = \frac{\int \varphi^* h \varphi d\tau}{\int \varphi^* \varphi d\tau} = \frac{\sum_p \sum_q c_p^* c_q \overbrace{\int \chi_p h \chi_q d\tau}^{h_{pq}}}{\sum_p \sum_q c_p^* c_q \underbrace{\int \chi_p \chi_q d\tau}_{S_{pq}}} = \epsilon(c_p^*, c_q)$$

The integrals over the AO's are abbreviated as indicated.
We then make use of the variational principle (without proof): By making the energy a minimum with respect to the coefficients c_p^* or equivalently c_q, the energy tends towards the lowest eigenvalue: $\epsilon_{min} \longrightarrow \epsilon_0$. Necessary conditions for a minimum are:

$$\frac{\partial \varepsilon}{\partial c_1^*} = \frac{\partial \varepsilon}{\partial c_2^*} = \ldots\ldots = \frac{\partial \varepsilon}{\partial c_p^*} = \ldots\ldots \frac{\partial \varepsilon}{\partial c_N^*} = 0$$

(We assume these conditions for our purposes also to be sufficient.)

We write:

$$\epsilon \left(\sum_p \sum_q c_p^* c_q S_{pq}\right) = \sum_p \sum_q c_p^* c_q h_{pq}$$

and differentiate with respect to c_p^*:

$$\frac{\partial \varepsilon}{\partial c_p^*} \left(\sum_p \sum_q c_p^* c_q S_{pq}\right) + \epsilon \left(\sum_q c_q S_{pq}\right) = \sum_q c_q h_{pq}$$

where p = 1,2 .... N. Setting the derivatives $\frac{\partial \varepsilon}{\partial c_p^*}$ equal to zero, we obtain the following equations:

$$\begin{cases} c_1(h_{11} - \epsilon S_{11}) + c_2(h_{12} - \epsilon S_{12}) + \ldots\ldots + c_N(h_{1N} - \epsilon S_{1N}) = 0 \\ c_1(h_{21} - \epsilon S_{21}) + c_2(h_{22} \quad \epsilon S_{22}) + \ldots\ldots + c_N(h_{2N} - \epsilon S_{2N}) = 0 \\ \quad \ldots\ldots\ldots \qquad \ldots\ldots\ldots \qquad \ldots\ldots\ldots \\ c_1(h_{N1} - \epsilon S_{N1}) + c_2(h_{N2} - \epsilon S_{N2}) + \ldots\ldots + c_N(h_{NN} - \epsilon S_{NN}) = 0 \end{cases}$$

This is a system of N linear homogeneous equations with N + 1 unknowns, namely N coefficients c_1 ... c_N, and the eigenvalue ϵ. These equations have non-trivial solutions only if the determinant of the coefficients vanishes:

$$\begin{vmatrix} (h_{11} - \epsilon S_{11}) & (h_{12} - \epsilon S_{12}) & \ldots\ldots & (h_{1N} - \epsilon S_{1N}) \\ \vdots & \vdots & & \vdots \\ (h_{N1} - \epsilon S_{N1}) & (h_{N2} - \epsilon S_{N2}) & \ldots\ldots & (h_{NN} - \epsilon S_{NN}) \end{vmatrix} = 0$$

This determinant is also called the secular determinant, its polynomial expansion the secular equation.

There will, in general, be N solutions for ε . To each eigenvalue ε_n there corresponds an eigenfunction φ_n. To obtain the coefficients c_{np} the condition of normalization must also be invoked:

$$\int \varphi_n^* \varphi_n d\tau \equiv \langle \varphi_n | \varphi_n \rangle = 1$$

The solutions then are:

eigenvalues	eigenvectors	coefficients
ε_1	$\varphi_1 = \sum_p c_{1p}\chi_p$	c_{11} , c_{12} .... c_{1N}
ε_2	$\varphi_2 = \sum_p c_{2p}\chi_p$	c_{21}, c_{22} .... c_{2N}
⋮	⋮	⋮
ε_N	$\varphi_N = \sum_p c_{Np}\chi_p$	c_{N1}, c_{N2} .... c_{NN}

Strictly speaking, from a variational point of view only the lowest solution is physically admissible. If the molecule of interest has certain elements of symmetry and the solutions transform according to different irreducible representations, then the lowest solution of each irreducible representation is admissible. In general, however, in the frame of the adopted crude approximations of the Hückel method, all solutions are considered meaningful.

2. Further simplifications

We write $h_{pp} \equiv \alpha_p$ and call it a coulomb integral

$h_{pq} \equiv \beta_{pq}$ and call it a resonance integral.

We neglect resonance integrals, except between nearest neighbors. We neglect overlap integrals; this corresponds to the zero differential overlap approximation.

<u>Example: Ethylene</u>

The secular equation is obtained as

$$\begin{vmatrix} \alpha-\epsilon & \beta \\ \beta & \alpha-\epsilon \end{vmatrix} = 0 \quad ; \quad (\alpha-\epsilon)^2 = \beta$$

and leads to the solutions $\begin{cases} \epsilon_1 = \alpha + \beta \\ \epsilon_2 = \alpha - \beta \end{cases}$

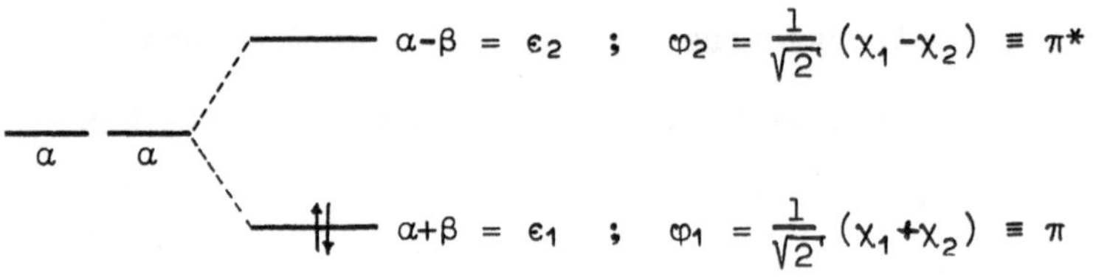

Figure 1 Hückel energy levels in ethylene

Physically, α may be assumed to correspond to the energy of an electron in a $2p_\pi$ orbital of an sp^2-hybridized carbon atom in its molecular surroundings; it is the negative of the corresponding atomic valence state ionization potential. β is a measure for the interaction between two such electrons on different carbon atoms, 1.34 Å apart. It may be calibrated empirically:

Thermochemical calibration of β:

i) For test molecules the enthalpy of formation ΔH is deduced from measured heats of combustion. It is compared with ΔH computed from additive increments for molecular fragments. The difference is attributed to a resonance energy (see page 12). The result is $\beta \sim 15$ to 20 kcal/mol.

ii) The barrier to internal rotation in ethylene, which is about 25 kcal/mol, is set equal to 2β. The result is $\beta \sim 12$ to 13 kcal/mol.

<u>Spectroscopic</u> calibration of β:

The longest-wavelength electronic transition in ethylene is (in part) composed of the $\pi \rightarrow \pi^*$ transition. It occurs roughly at 180 nm. $\lambda^{-1} = 55'000\ \text{cm}^{-1}$

$$h\nu = \Delta E = \epsilon_2 - \epsilon_1 = 2\beta \approx 7\ \text{eV} \quad ; \quad \beta \approx 3.5\ \text{eV}$$

<u>We note:</u>

Thermochemical predictions require thermochemical calibrations of β on a test molecule;
spectroscopic predictions require spectroscopic calibrations.

<u>Exercise:</u>

Butadiene; butadiene with symmetry orbitals; analogy with solutions of the free electron in a box.

<u>Example: Benzene</u>

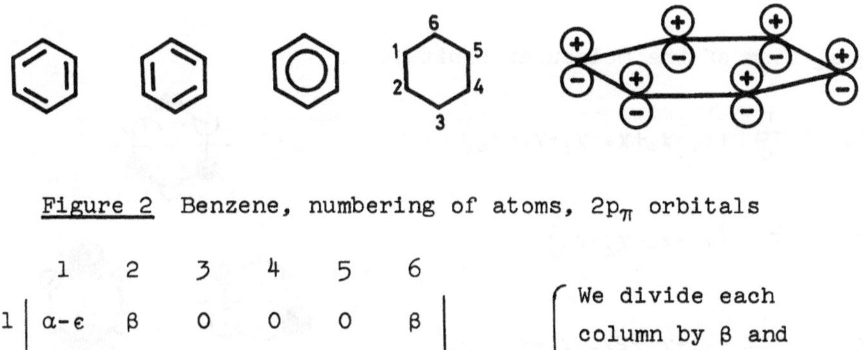

<u>Figure 2</u> Benzene, numbering of atoms, $2p_\pi$ orbitals

$$\begin{array}{c|cccccc|}
 & 1 & 2 & 3 & 4 & 5 & 6 \\
1 & \alpha-\epsilon & \beta & 0 & 0 & 0 & \beta \\
2 & \beta & \alpha-\epsilon & \beta & 0 & 0 & 0 \\
3 & 0 & \beta & \alpha-\epsilon & \beta & 0 & 0 \\
4 & 0 & 0 & \beta & \alpha-\epsilon & \beta & 0 \\
5 & 0 & 0 & 0 & \beta & \alpha-\epsilon & \beta \\
6 & \beta & 0 & 0 & 0 & \beta & \alpha-\epsilon
\end{array} = 0$$

We divide each column by β and abbreviate

$$\frac{\alpha - \epsilon}{\beta} \equiv x \quad ;$$

We thus get the secular equation in the form:

$$\begin{vmatrix} x & 1 & 0 & 0 & 0 & 1 \\ 1 & x & 1 & 0 & 0 & 0 \\ 0 & 1 & x & 1 & 0 & 0 \\ 0 & 0 & 1 & x & 1 & 0 \\ 0 & 0 & 0 & 1 & x & 1 \\ 1 & 0 & 0 & 0 & 1 & x \end{vmatrix} = 0 \qquad \left\{ \begin{array}{l} \text{One obtains the solutions:} \\ x_k = -2\cos\frac{2k\pi}{6}, \text{ where} \\ k = 0, \pm 1, \pm 2, 3 \end{array} \right.$$

We then obtain the following energy level scheme and eigenfunctions

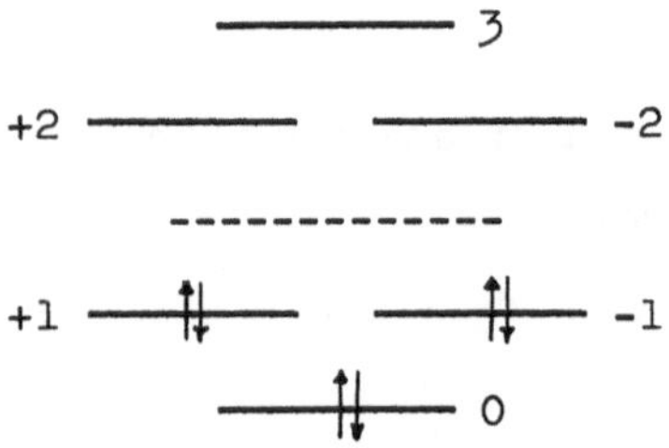

$\varepsilon_3 = \alpha - 2\beta \qquad \varphi_3$

$\varepsilon_{\pm 2} = \alpha - \beta \qquad \varphi_{+2}, \varphi_{-2}$

$\varepsilon_{\pm 1} = \alpha + \beta \qquad \varphi_{+1} \quad \varphi_{-1}$

$\varepsilon_0 = \alpha + 2\beta \qquad \varphi_0$

Real form of the molecular orbitals:

$$\varphi_3 = \frac{1}{\sqrt{6}} \{\chi_1 - \chi_2 + \chi_3 - \chi_4 + \chi_5 - \chi_6\}$$

$$\varphi_{-2} = \frac{1}{2} \{\chi_1 - \chi_2 + \chi_4 - \chi_5\}$$

$$\varphi_{+2} = \frac{1}{\sqrt{12}} \{\chi_1 + \chi_2 - 2\chi_3 + \chi_4 + \chi_5 - 2\chi_6\}$$

$$\varphi_{-1} = \frac{1}{2} \{\chi_1 + \chi_2 - \chi_4 - \chi_5\}$$

$$\varphi_{+1} = \frac{1}{\sqrt{12}} \{\chi_1 - \chi_2 - 2\chi_3 - \chi_4 + \chi_5 + 2\chi_6\}$$

$$\varphi_0 = \frac{1}{\sqrt{6}} \{\chi_1 + \chi_2 + \chi_3 + \chi_4 + \chi_5 + \chi_6\}$$

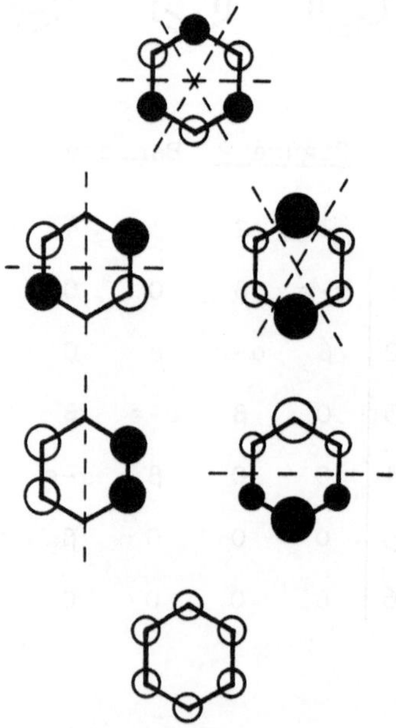

Figure 3 Real benzene MO's

Complex form of the molecular orbitals:

$$b = \frac{1}{\sqrt{6}} \sum_{p=1}^{N} (-1)^p \chi_p = \frac{1}{\sqrt{6}} \sum_p \exp(+3p\cdot 2\pi i/6) \cdot \chi_p = \varphi_3$$

$$e_{\pm 2} = \frac{1}{\sqrt{6}} \sum_{p=1}^{N} \exp(\pm 2p\cdot 2\pi i/6) \cdot \chi_p$$

$$e_{\pm 1} = \frac{1}{\sqrt{6}} \sum_p \exp(\pm p\cdot 2\pi i/6) \cdot \chi_p$$

$$a = \frac{1}{\sqrt{6}} \sum_p \chi_p = \varphi_0$$

These complex orbitals are symmetry orbitals of the subgroup C_6 of D_{6h}.

The relations between the degenerate real and complex solutions are

$$\frac{1}{\sqrt{2}} (e_{+1} + e_{-1}) = \varphi_{+1} \; ; \quad \frac{1}{\sqrt{2}} (e_{+2} + e_{-2}) = \varphi_{+2}$$

$$\frac{1}{i\sqrt{2}} (e_{+1} - e_{-1}) = \varphi_{-1} \; ; \quad \frac{1}{i\sqrt{2}} (e_{+2} - e_{-2}) = \varphi_{-2}$$

Later on we will see that it is more convenient to use complex orbitals than the real ones. Physically they are of course equivalent, for any linear combination of two eigenfunctions belonging to the same eigenvalue is again an eigenfunction to that eigenvalue.

3. Some important definitions

Atomic density: (Total atomic population)

$$Q_r = \sum_{i=1}^{N} b_i c_{ir}^* c_{ir} \; ; \quad \left\{ \begin{array}{l} \text{occupation number} \\ b_i = 0,\, 1,\, 2 \end{array} \right.$$

It is a measure of the amount of π electrons on atom p. For all alternant hydrocarbons in their ground state $Q_r = 1$ on all atoms.

<u>Bond order:</u> $$P_{rs} = \sum_{i=1}^{N} b_i c_{ir}^{*} c_{is} \quad ; \quad b_i = 0,\ 1,\ 2$$

Within the frame of Hückel theory this is the first-order density matrix.

<u>Energy of a configuration:</u> It is the sum of the one-electron energies of the π electrons in the system of interest.

$$E = \sum_{i=1}^{N} b_i \epsilon_i$$

The lowest configuration possible is the ground configuration. In it the lowest one-electron levels are all doubly occupied. It is an approximation to the many-electron ground state. For benzene we have

$$E_G = 2\epsilon_0 + 2\epsilon_{+1} + 2\epsilon_{-1} = 6\alpha + 8\beta$$

<u>Resonance energy:</u> It corresponds to the difference

$$\{E_G\ (\pi \text{ electrons completely delocalized}) - E_G\ (\pi \text{ electrons localized in double bonds})\}$$

For benzene this may be visualized as }

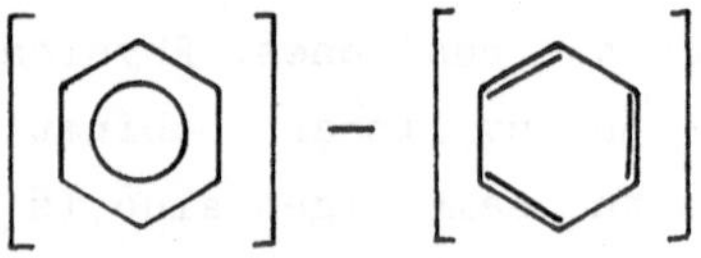

and amounts to

$$E_{Resonance} = (6\alpha + 8\beta) - (6\alpha + 6\beta) = 2\beta$$

<u>Alternant hydrocarbons:</u> They may be divided into <u>alternate</u> nonneighboring starred and unstarred carbon atoms

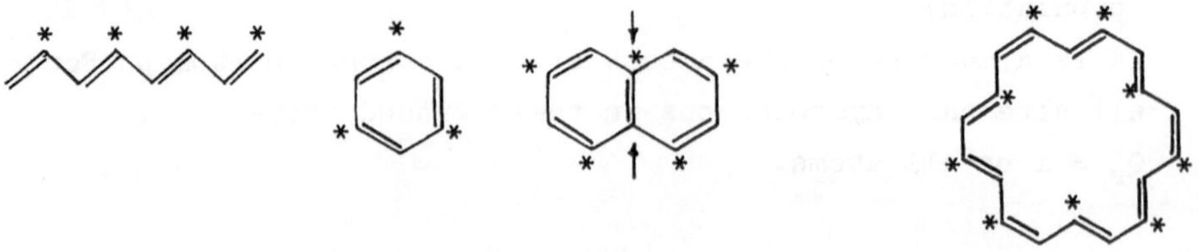

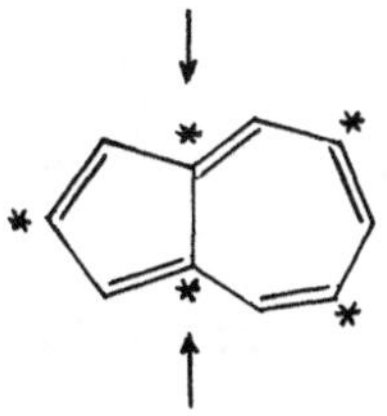

In nonalternant hydrocarbons there occurs at least one bond between two starred (or two unstarred) atoms.

In alternant hydrocarbons energy levels are pairwise equidistantly spaced with respect to α (provided overlap is neglected):

$$\epsilon_i = \alpha + x_i\beta \;, \qquad \varphi_i = \sum_p^{*} c_{ip}\chi_p + \sum_{p'}^{o} c_{ip'}\chi_{p'}$$

$$\epsilon_{(N+1-i)} = \alpha - x_i\beta \;, \qquad \varphi_{(N+1-i)} = \sum_p^{*} c_{ip}\chi_p - \sum_{p'}^{o} c_{ip'}\chi_{p'}$$

The corresponding eigenfunctions are characterized by the fact that for starred atoms the coefficients are the same and for unstarred atoms they are of equal absolute value, but opposite sign. The absolute designation of an atom as "starred" or "unstarred" is of course arbitrary, but such is also the absolute sign of the molecular orbital.

The so-called "pairing" of electronic states in alternant hydrocarbons has its physical consequences. For instance, it follows from the Hückel model that in the radical anion and radical cation of an alternant hydrocarbon the spin distribution should be identical. The experimental proton hyperfine splittings in the electron spin resonance spectrum of such an anion and cation are indeed remarkably similar [4], provided they are unsubstituted.

The inclusion of heteroatoms:

Much has been written and said on this topic.

In general one may write:

$$\begin{cases} \alpha_X = \alpha_C + h_X\,\beta_{CC} \\ \beta_{CX} = k_{CX}\,\beta_{CC} \end{cases}$$

Examples [3c] : 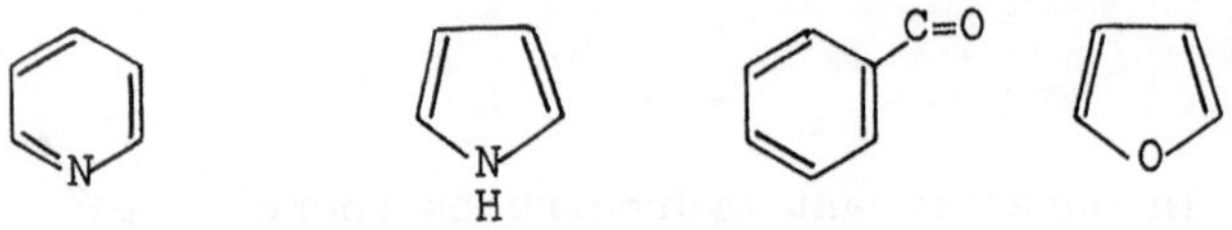

$h_{\dot{N}} = 0.5 - 1.0$ $\quad h_{\ddot{N}} = 1.5$ $\quad h_{\dot{O}} = 1.$ $\quad h_{\ddot{O}} = 2.$

$k_{CN} = 1.$ $\quad k_{CN} = 0.8$ $\quad k_{CO} = 1.$ $\quad k_{CO} = 0.8$

In general one may assume $\beta_{CX} \approx \mu \cdot S_{CX}$, μ being a proportionality constant. Caution must be exercised in applying this relation to 3d.-row elements.

<u>Some useful relations:</u>

From $E = \sum_i b_i \epsilon_i$ follows (without proof)

$$\frac{\partial E}{\partial \alpha_s} = Q_s \quad , \quad \frac{1}{2}\frac{\partial E}{\partial \beta_{sr}} = P_{sr}$$

Furthermore one defines:

$$\frac{\partial Q_s}{\partial \alpha_q} \equiv \Pi_{s,q} \quad \text{atom-atom polarizability} \quad = \frac{\partial^2 E}{\partial \alpha_s \partial \alpha_q}$$

$$\frac{\partial P_{sr}}{\partial \beta_{qt}} \equiv \Pi_{sr,qt} \quad \text{bond-bond polarizability} \quad = \frac{1}{2}\frac{\partial^2 E}{\partial \beta_{sr} \partial \beta_{qt}}$$

These quantities prove useful in applying perturbation theory to the Hückel procedure, as they are related to the first and higher derivatives of the energy with respect to basic quantities, and as they may be computed from eigenvalues ϵ_i and eigenvectors c_{ip}. The pertinent formulae are to be found in ref. 3f, for instance, and both atom-atom and bond-bond polarizabilities are tabulated for a large variety of hydrocarbons in ref. 3e. However, due to increased computer facilities, the use of perturbation techniques in the frame of Hückel theory has lost some importance in recent years.

III. Many-electron theory of π-electrons

We now explicitly consider the interaction between π-electrons [5]. Our Hamiltonian has the form

$$\mathcal{H}^{\pi} = \sum_{\mu}^{(\pi)} h_{\text{Core}}(\mu) + \sum_{\mu>\nu}^{(\pi)} \frac{e^2}{r_{\mu\nu}}$$

and we no longer consider each π electron to merely move in the inaccurately defined average field of all the others, but rather to depend more explicitly on their relative positions.

We make the following formal assumptions:

a) Each electron μ may in zeroth order of approximation be described by a spatial one-electron function or orbital (in our case a molecular orbital, MO) $\varphi(\mu)$.

b) Each electron is, with respect to all the others, in a definite spin-state $\alpha(\mu)$ or $\beta(\mu)$. We thus associate with every electron a spinorbital $\varphi(\mu)\cdot\alpha(\mu)$ or $\varphi(\mu)\cdot\beta(\mu)$. In the following sections we will often abbreviate $\varphi\alpha \equiv \varphi$, $\varphi\beta \equiv \bar{\varphi}$.

c) Any many-electron function must be antisymmetric with respect to the exchange of two electrons, as required by the Pauli-principle.

   Consequently such a function is best represented by a Slater determinant or by a linear combination of Slater determinants, each such Slater determinant being an antisymmetrized product of one-electron functions, i.e. of spinorbitals.

1. Ethylene as two-electron problem

We denote the two carbon atoms by a and b and invoke the σ-π separability. We thus have a pseudo-two-electron problem. Starting from our previous one-electron energy level scheme, we can construct 6 configurations consistent with the Pauli-principle.

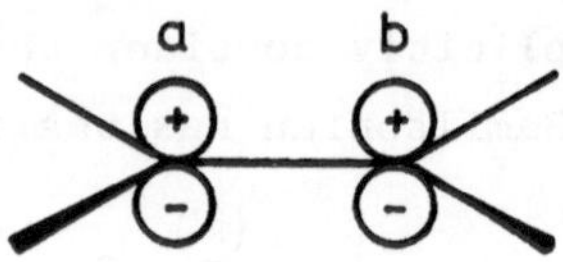

Figure 4

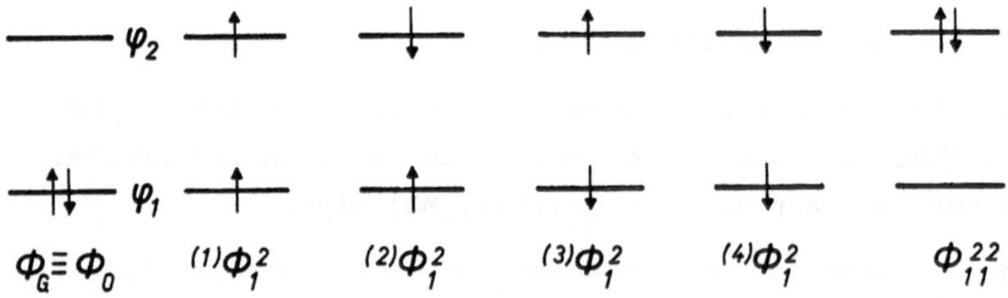

Figure 5

Of course: $\varphi_1 = \dfrac{1}{\sqrt{2(1+S_{ab})}} \left\{\chi_a + \chi_b\right\}$

$$\varphi_2 = \frac{1}{\sqrt{2(1-S_{ab})}} \left\{\chi_a - \chi_b\right\}$$

and the ground configuration is written:

$$\Phi_G = \frac{1}{\sqrt{2}} \begin{vmatrix} \varphi_1\alpha(1) & \varphi_1\beta(1) \\ \varphi_1\alpha(2) & \varphi_1\beta(2) \end{vmatrix} = \frac{1}{\sqrt{2}}\left\{\varphi_1\alpha(1)\varphi_1\beta(2) - \varphi_1\beta(1)\varphi_1\alpha(2)\right\}$$

$$= \frac{1}{\sqrt{2}}\varphi_1(1)\varphi_1(2)\left\{\alpha(1)\beta(2) - \beta(1)\alpha(2)\right\}$$

We abbreviate $\Phi_G \equiv |\varphi_1\bar{\varphi}_1|$, namely as the diagonal part of the Slater determinant, omitting the normalization factor. The excited configurations are consequently written:

All the singly excited functions are degenerate to zeroth order.

$$\left.\begin{aligned} {}^{(1)}\Phi_1^2 &= |\varphi_1\varphi_2| \\ {}^{(2)}\Phi_1^2 &= |\varphi_1\overline{\varphi}_2| \\ {}^{(3)}\Phi_1^2 &= |\overline{\varphi}_1\varphi_2| \\ {}^{(4)}\Phi_1^2 &= |\overline{\varphi}_1\overline{\varphi}_2| \end{aligned}\right\} \quad {}^{(1)}E_1^2 = {}^{(2)}E_1^2 = {}^{(3)}E_1^2 = {}^{(4)}E_1^2 = \alpha + \beta + \alpha - \beta = 2\alpha$$

(in the Hückel approximation)

$$\Phi_{11}^{22} = |\varphi_2\overline{\varphi}_2|$$

We assume that a true description of the electrons which we consider will not be given by <u>one</u> configurational function, but by a linear combination of them:

$$\psi = \sum_{\lambda} A_\lambda \Phi_\lambda$$

To determine the correct expansion coefficients A_λ we apply the variational principle in formally exactly the same way as we did in the one-electron case. This leads to a secular equation

$$\left|\mathcal{H}^{\pi}_{\lambda\lambda'} - E\, S_{\lambda\lambda'}\right| = 0$$

where the indices λ and λ' run over all configurations. The dimension of our secular determinant is equal to the number of configurations of interest.

$$\mathcal{H}^{\pi}_{\lambda\lambda'} \equiv \langle\Phi_\lambda|\mathcal{H}^{\pi}|\Phi_{\lambda'}\rangle \equiv \int \Phi_\lambda^* \mathcal{H}^{\pi} \Phi_{\lambda'}\, d\tau$$

$$S_{\lambda\lambda'} \equiv \langle\Phi_\lambda|\Phi_{\lambda'}\rangle \equiv \int \Phi_\lambda^* \Phi_{\lambda'}\, d\tau = \delta_{\lambda\lambda'}$$

Now, before we compute the matrix elements of $\mathcal{H}^{\pi}$ in the basis of the configurational functions, let us, however, consider the following:

As the operators S^2 and S_Z for the total spin commute with the Hamiltonian (in absence of spin-orbit effects)

$$[\mathcal{H}^\pi, S^2] = 0 \quad , \quad [\mathcal{H}^\pi, S_Z] = 0$$

matrix elements of $\mathcal{H}^\pi$ will vanish between configurational functions which are eigenfunctions of S^2 and S_Z with different eigenvalues [6].

Of course $\vec{S} = \vec{s}_1 + \vec{s}_2$ and $S_Z = s_{1Z} + s_{2Z}$

$$S^2 = s_1^2 + s_2^2 + 2\vec{s}_1 \cdot \vec{s}_2$$

$$= s_1^2 + s_2^2 + s_1^+ \cdot s_2^- + s_1^- \cdot s_2^+ + 2s_{1Z} \cdot s_{2Z}$$

where $s_1^+ \equiv s_{1x} + is_{1y}$, $s_2^- \equiv s_{2x} - is_{2y}$; etc.

We remember that

$$s_{1Z}\alpha(1) = \frac{\hbar}{2}\,\alpha(1) \,, \qquad s_{1Z}\beta(1) = -\frac{\hbar}{2}\,\beta(1)$$

$$s_1^+\,\alpha(1) = 0 \,, \qquad s_1^+\,\beta(1) = \hbar\,\alpha(1)$$

$$s_1^-\,\alpha(1) = \hbar\,\beta(1) \,, \qquad s_1^-\,\beta(1) = 0$$

$$s_1^2\,\alpha(1) = \frac{3}{4}\,\hbar^2\alpha(1), \qquad s_1^2\,\beta(1) = \frac{3}{4}\,\hbar^2\beta(1)$$

etc.,

consequently:

$$S_Z\,\Phi_G = 0 \qquad \text{and} \qquad S^2\,\Phi_G = 0$$

$$S_Z\,{}^{(1)}\Phi_1^2 = 1\cdot\hbar\cdot{}^{(1)}\Phi_1^2 \qquad S^2\,{}^{(1)}\Phi_1^2 = 2\cdot\hbar^2\cdot{}^{(1)}\Phi_1^2$$

$$S_Z\,{}^{(2)}\Phi_1^2 = 0 \qquad S^2\,{}^{(2)}\Phi_1^2 = \hbar^2\left\{{}^{(2)}\Phi_1^2 + {}^{(3)}\Phi_1^2\right\}$$

$$S_Z\,{}^{(3)}\Phi_1^2 = 0 \qquad S^2\,{}^{(3)}\Phi_1^2 = \hbar^2\left\{{}^{(2)}\Phi_1^2 + {}^{(3)}\Phi_1^2\right\}$$

$$S_Z\,{}^{(4)}\Phi_1^2 = -1\cdot\hbar\cdot{}^{(4)}\Phi_1^2 \qquad S^2\,{}^{(4)}\Phi_1^2 = 2\cdot\hbar^2\cdot{}^{(4)}\Phi_1^2$$

$$S_Z\,\Phi_{11}^{22} = 0 \qquad S^2\,\Phi_{11}^{22} = 0$$

<u>Exercise:</u> 1) Derive the above relations
2) Prove that $|\bar{\varphi}_1 \varphi_2| \;=\; -|\varphi_2 \bar{\varphi}_1|$

Our above functions are already eigenfunctions of S_Z. We are now in a position to write eigenfunctions of both S_Z and S^2. We designate these functions by ${}^{(2S+1)}_{M_S}\Phi_i^j$, where S and M_S are total spin and z-component-of-spin quantum numbers.

<u>Singlets</u>	<u>Triplets</u>
${}^1_0\Phi_G = \lvert\varphi_1\bar{\varphi}_1\rvert$	${}^3_{-1}\Phi_1^2 = \lvert\bar{\varphi}_1\bar{\varphi}_2\rvert$
${}^1_0\Phi_1^2 = \frac{1}{\sqrt{2}}\left\{\lvert\varphi_1\bar{\varphi}_2\rvert + \lvert\varphi_2\bar{\varphi}_1\rvert\right\}$	${}^3_0\Phi_1^2 = \frac{1}{\sqrt{2}}\left\{\lvert\varphi_1\bar{\varphi}_2\rvert - \lvert\varphi_2\bar{\varphi}_1\rvert\right\}$
${}^1_0\Phi_{11}^{22} = \lvert\varphi_2\bar{\varphi}_2\rvert$	${}^3_1\Phi_1^2 = \lvert\varphi_1\varphi_2\rvert$

We now make use of these 6 functions to describe the electronic states of ethylene. As mentioned, matrix elements of $\mathcal{H}^\pi$ will vanish between functions of different S or M_S values. So the triplet functions, as they stand, are already solutions.

We now compute nonvanishing matrix elements of

$$\mathcal{H}^\pi = h_{core}(1) + h_{core}(2) + \frac{e^2}{r_{12}}$$

between the singlet functions. The matrix elements are expanded into one and two-electron integrals, as indicated. Only the integration over the spatial variables is explicitly mentioned. Of course, the orthogonality of $\alpha(\mu)$ und $\beta(\mu)$ must also be observed. $h_{core}(\mu)$ is just written $h(\mu)$, where $\mu = 1,2$.

$$\langle {}^1_0\Phi_G|\mathcal{H}^\pi|{}^1_0\Phi_G\rangle = \int |\varphi_1\bar{\varphi}_1|^* h(1)\,|\varphi_1\bar{\varphi}_1|\,d\tau_1 d\tau_2 + \text{id. for } h(2) + $$

$$+ \int |\varphi_1\bar{\varphi}_1|^* \frac{e^2}{r_{12}}\,|\varphi_1\bar{\varphi}_1|\,d\tau_1 d\tau_2$$

Upon expansion of the determinants:

$$= 2\int \varphi_1^* h \varphi_1 d\tau_1 + \int \varphi_1^*(1)\varphi_1^*(2) \frac{e^2}{r_{12}} \varphi_1(1)\varphi_1(2) d\tau_1 d\tau_2$$

For this expression we use the abbreviations

$$\equiv 2\langle \varphi_1 | h | \varphi_1 \rangle + \langle \varphi_1 \varphi_1 | \varphi_1 \varphi_1 \rangle$$

or equivalently

$$\equiv 2h_{11} + \langle 11|11\rangle$$

Similarly:

$$\langle {}^1_0\Phi_G | \mathcal{H}^\pi | {}^1_0\Phi_1^2 \rangle = \frac{1}{\sqrt{2}} \int |\varphi_1 \bar{\varphi}_1|^* h(1) \left\{ |\varphi_1 \bar{\varphi}_2| + |\varphi_2 \bar{\varphi}_1| \right\} d\tau_1 d\tau_2 + \text{id. } h(2)$$

$$+ \frac{1}{\sqrt{2}} \int |\varphi_1 \bar{\varphi}_1|^* \frac{e^2}{r_{12}} \left\{ |\varphi_1 \bar{\varphi}_2| + |\varphi_2 \bar{\varphi}_1| \right\} d\tau_1 d\tau_2$$

$$= \sqrt{2} \int \varphi_1^* h \varphi_2 d\tau_1 + \sqrt{2} \int \varphi_1^*(1)\varphi_1^*(2) \frac{e^2}{r_{12}} \varphi_1(1)\varphi_2(2) d\tau_1 d\tau_2$$

$$\equiv \sqrt{2} \left\{ \langle \varphi_1 | h | \varphi_2 \rangle + \langle \varphi_1 \varphi_1 | \varphi_1 \varphi_2 \rangle \right\}$$

$$\equiv \sqrt{2} \left\{ h_{12} + \langle 11|12\rangle \right\} = 0$$

(To ascertain this result, the integrals over molecular orbitals must be expanded into integrals over atomic orbitals; see below.)

In the same way we find:

$$\langle {}^1_0\Phi_G | \mathcal{H}^\pi | {}^1_0\Phi_{11}^{22} \rangle = \langle 11|22\rangle$$

$$\langle {}^1_0\Phi_1^2 | \mathcal{H}^\pi | {}^1_0\Phi_1^2 \rangle = h_{11} + h_{22} + \langle 12|12\rangle + \langle 12|21\rangle$$

$$\langle {}^1_0\Phi_1^2 | \mathcal{H}^\pi | {}^1_0\Phi_{11}^{22} \rangle = \sqrt{2} \left\{ h_{21} + \langle 22|21\rangle \right\} = 0$$ to ascertain this, see below)

$$\langle {}^1_0\Phi_{11}^{22} | \mathcal{H}^\pi | {}^1_0\Phi_{11}^{22} \rangle = 2h_{22} + \langle 22|22\rangle$$

From these results the singlet secular equation is obtained.

$$
\begin{array}{c|cc|c|}
 & {}^1_0\Phi_G & {}^1_0\Phi_{11}^{22} & {}^1_0\Phi_1^2 \\
\hline
{}^1_0\Phi_G & 2h_{11}+\langle 11|11\rangle-E & \langle 11|22\rangle & 0 \\
{}^1_0\Phi_{11}^{22} & \langle 11|22\rangle & 2h_{22}+\langle 22|22\rangle-E & 0 \\
\hline
{}^1_0\Phi_1^2 & 0 & 0 & h_{11}+h_{22}+\langle 12|12\rangle+\langle 12|21\rangle-E \\
\hline
\end{array} = 0
$$

We notice that the equation factorizes into a (2×2) equation, connecting the ground configuration with the doubly excited configuration, and into a (1×1) equation, of which ${}^1_0\Phi_1^2$ is an eigenfunction with eigenvalue

$$^1E_1 = h_{11} + h_{22} + \langle 12|12\rangle + \langle 12|21\rangle$$

Consequently, to summarize, our singlet eigenfunctions have the form:

$$^1\psi_0 \equiv \psi_G = A_I\, \Phi_G + A_{II}\, \Phi_{11}^{22}$$

$$^1\psi_1 = {}^1_0\Phi_1^2$$

$$^1\psi_2 = A_I\, \Phi_{11}^{22} - A_{II}\, \Phi_G$$

The coefficients A_I and A_{II} have to be numerically determined, but evidently $A_I \gg A_{II}$.

The triplet secular equation has no off-diagonal elements for reasons already discussed. Indeed we have

$$^3E_1 = \langle {}^3_{-1}\Phi_1^2 | \mathcal{H}^\pi | {}^3_{-1}\Phi_1^2 \rangle = \langle {}^3_0\Phi_1^2 | \mathcal{H}^\pi | {}^3_0\Phi_1^2 \rangle = \langle {}^3_{+1}\Phi_1^2 | \mathcal{H}^\pi | {}^3_{+1}\Phi_1^2 \rangle$$

$$= h_{11} + h_{22} + \langle 12|12\rangle - \langle 12|21\rangle$$

The three triplet eigenfunctions are given on page 19.

Figure 6 shows at left the relative energies of the various configurations, neglecting electron interaction, and just summing the one-electron energies. Note the degeneracy between $^1\Phi_1^2$ and $^3\Phi_1^2$. At right, in a somewhat arbitrary scale, are the results with electron interaction. A striking feature is the energy splitting between singlet and triplet ψ_1.

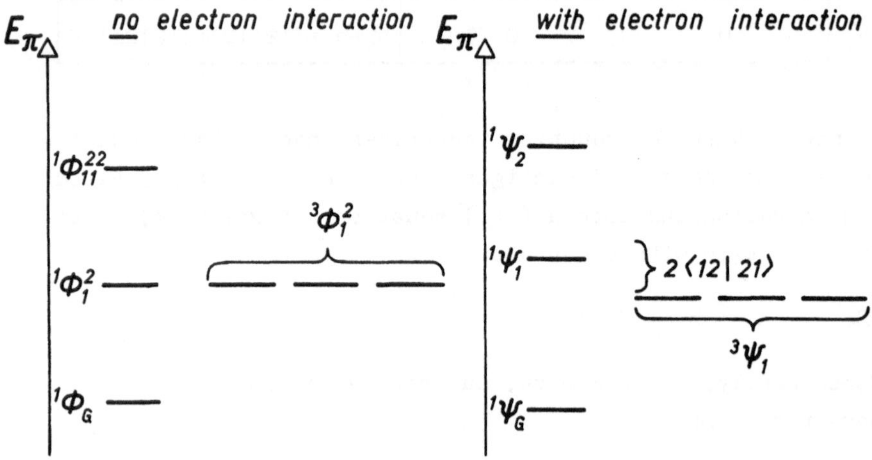

Figure 6

The use of spatial symmetry:

(The reader is here assumed to be familiar with elementary group theory)

The molecule ethylene has the symmetry D_{2h}.

The MO φ_1 transforms according to the irreducible representation b_{1u}.

The MO φ_2 transforms according to the irreducible representation b_{2g}.

(We use small lettering to characterize one-electron states and capital letters to label many-electron states.)

Consequently	Φ_G	transforms like	A_g	=	$b_{1u} \otimes b_{1u}$
	Φ_1^2	" "	B_{3u}	=	$b_{1u} \otimes b_{2g}$
	Φ_{11}^{22}	" "	A_g	=	$b_{2g} \otimes b_{2g}$

As $\mathcal{K}^{\pi}$ transforms like A_g, matrix elements between configurational functions belonging to different irreducible representations vanish. This explains the factoring of the singlet secular equation.

<u>The evaluation of integrals:</u>

For numerical computations integrals over molecular orbitals must of course be further expanded into integrals over atomic orbitals. In this sense we obtain [5]:

a) Two-electron integrals:

$$\langle 11|11\rangle \equiv J_{11} = \frac{\langle \overset{(1)(2)}{aa}|\overset{(1)(2)}{aa}\rangle + \langle ab|ab\rangle + 2\langle aa|bb\rangle + 4\langle aa|ab\rangle}{2(1+S_{ab})^2}$$

$$\langle 22|22\rangle \equiv J_{22} = \frac{\langle aa|aa\rangle + \langle ab|ab\rangle + 2\langle aa|bb\rangle - 4\langle \overset{(1)(2)}{aa}|\overset{(1)(2)}{ab}\rangle}{2(1-S_{ab})^2}$$

$$\langle 12|12\rangle \equiv J_{12} = \frac{\langle aa|aa\rangle + \langle ab|ab\rangle - 2\langle aa|bb\rangle}{2(1-S_{ab}^2)}$$

$$\langle 12|21\rangle \equiv K_{12} = \frac{\langle aa|aa\rangle - \langle ab|ab\rangle}{2(1-S_{ab}^2)}$$

Consistent with our previous notation

$$\langle ab|ab\rangle \equiv \int \chi_a\overset{\mu}{(1)}\chi_b\overset{\nu}{(2)}\,\frac{e^2}{r_{12}}\,\chi_a(1)\chi_b(2)d\tau_1 d\tau_2 \quad \Big(\equiv (aa|bb)\Big)$$

$$\langle aa|bb\rangle \equiv \int \chi_a(1)\chi_a(2)\,\frac{e^2}{r_{12}}\,\chi_b(1)\chi_b(2)d\tau_1 d\tau_2 \quad \Big(\equiv (ab|ab)\Big)$$

The round-bracket notation is used in [5]. These integrals are either calculated <u>accurately</u> or evaluated <u>semiempirically</u>. The semiempirical evaluation is important in our context, i.e. in large molecules, and will therefore be dealt with:

- Zero differential overlap approximation: Wherever the product $\chi_a(1)\chi_b(1)$ occurs it is neglected, unless $\chi_a = \chi_b$. With this approximation $S_{ab} = 0$; $\langle aa|bb\rangle = 0$; $\langle aa|ab\rangle = 0$.

- We approximate one-center two-electron integrals as
  $\langle aa|aa\rangle \approx I_a - EA_a$

  where I_a is the potential of ionization of an electron located in the $2p_\pi$ orbital χ_a on the sp^2 hybridized carbon atom a in its molecular surroundings, and EA_a is the corresponding valence state electron affinity. The above equation may be visualized as the transfer of an electron from one carbon atom in the given valence state to another identical one infinitely far away. The energy required to carry out that transfer is, on one hand, I_a-EA_a, on the other it may be viewed simply as the work required to overcome the repulsion energy $\langle aa|aa\rangle$.

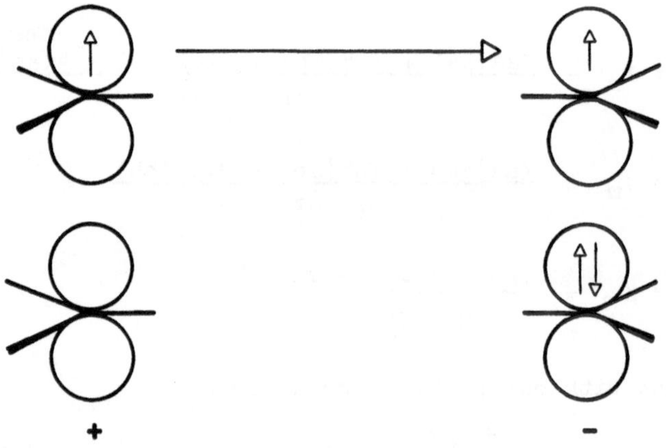

Figure 7

Two-center two-electron integrals of the type $\langle ab|ab\rangle$ may be semiempirically approximated by an electrostatic model (for details, see next section).

b) One-electron core integrals:

$$h_{11} \equiv \langle \varphi_1|h|\varphi_1\rangle = \frac{\langle a|h|a\rangle + \langle b|h|b\rangle + \langle a|h|b\rangle + \langle b|h|a\rangle}{2(1+S_{ab})}$$

$$= \frac{\langle a|h|a\rangle + \langle a|h|b\rangle}{1+S_{ab}}$$

We consider the different terms contained in the core operator $h(1) \equiv h(\mu)$:

$$h(1) = T(1) + \underbrace{U_a(1) + U_b(1)}_{\substack{\text{Interaction of the} \\ \text{electron with the} \\ \text{positive core of} \\ \text{carbon atoms a and b}}} + \underbrace{\left\{U'_{H_1}(1) + U'_{H_2}(1) + U'_{H_3}(1) + U'_{H_4}(1)\right\}}_{\substack{\text{Relatively small interaction of} \\ \text{the electron with the neutral} \\ \text{hydrogen atoms. These terms are} \\ \text{in general considered negligible}}}$$

The matrix element of the sum of the first two terms between identical functions χ_a may be viewed as the negative of the "valence state" ionization potential I_a:

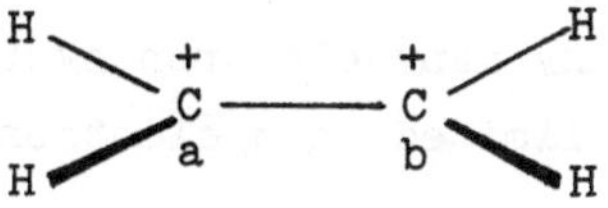

Figure 8

$$\langle \chi_a(1)|T(1) + U_a(1)|\chi_a(1)\rangle \equiv \langle a|T + U_a|a\rangle \approx -I_a$$

The matrix element of U_b between identical functions χ_a may be considered as the interaction of an electron in χ_a with the positive hole created by the vacancy of an electron in χ_b, and set approximately equal to:

$$\langle \chi_a(1)|U_b(1)|\chi_a(1)\rangle \equiv \langle a|U_b|a\rangle \approx - \langle \chi_a(1)\chi_b(2)|\frac{e^2}{r_{12}}|\chi_a(1)\chi_b(2)\rangle$$

$$\approx - \langle ab|ab\rangle$$

Thus:

$$\alpha_a \equiv \langle a|h|a\rangle \approx -I_a - \langle ab|ab\rangle$$

The matrix element of h between orbitals on different centers, χ_a and χ_b, is difficult to interpret physically term-by-term. For larger π electron systems (see also section III.3) the core resonance integrals

$$\beta_{ab} \equiv \langle a|h|b\rangle$$

are generally calibrated on a test molecule. The quantities α_a and β_{ab}, as described here, are not to be confused with the more crudely defined coulomb and resonance integrals of Hückel theory, in spite of the similarity.

2. The configuration interaction (CI) procedure

The procedure outlined here is, in principle, applicable to any many-electron system. Our formulation is consequently not limited to π electrons.

We consider a molecular system with 2N electrons, described by a Hamiltonian

$$\mathcal{H} = \sum_{\mu} h(\mu) + \sum_{\mu>\nu}\sum v(\mu,\nu) \quad , \quad \text{where } v(\mu,\nu) \equiv \frac{e^2}{r_{\mu\nu}}$$

The wavefunctions for the many-electron system will be linear combinations of Slater determinants, which in turn are defined as antisymmetrized products of spinorbitals (spin-MO's). The MO's are written as linear combinations of basis orbitals (AO's):

$$\varphi_i = \sum_p c_{ip}\chi_p$$

We assume that the MO's are ordered with respect to an energy criterion. They are, in the case of π electrons, for instance, Hückel MO's ordered in the sequence of increasing energy. We are thus able to define a

ground configuration $\Phi_G = |\varphi_1\bar{\varphi}_1 \ldots \varphi_i\bar{\varphi}_i \ldots \varphi_N\bar{\varphi}_N|$

singly excited configurations,

singlet $${}^1\Phi_i^k = \frac{1}{\sqrt{2}}\left\{|\varphi_1\bar{\varphi}_1 \ldots \varphi_i\bar{\varphi}_k \ldots \varphi_N\bar{\varphi}_N| + |\varphi_1\bar{\varphi}_1 \ldots \varphi_k\bar{\varphi}_i \ldots \varphi_N\bar{\varphi}_N|\right\}$$

$$\text{triplet}\begin{cases} {}^{3}_{-1}\Phi_{i}^{k} = |\varphi_1\bar{\varphi}_1 \ldots \bar{\varphi}_i\bar{\varphi}_k \ldots \varphi_N\bar{\varphi}_N| \\ {}^{3}_{0}\Phi_{i}^{k} = \frac{1}{\sqrt{2}}\left\{|\varphi_1\bar{\varphi}_1 \ldots \varphi_i\bar{\varphi}_k \ldots \varphi_N\bar{\varphi}_N| - |\varphi_1\bar{\varphi}_1 \ldots \varphi_k\bar{\varphi}_i \ldots \varphi_N\bar{\varphi}_N|\right\} \\ {}^{3}_{+1}\Phi_{i}^{k} = |\varphi_1\bar{\varphi}_1 \ldots \varphi_i\varphi_k \ldots \varphi_N\bar{\varphi}_N| \end{cases}$$

doubly excited $\Phi_{ij}^{k\ell}$, triply excited configurations $\Phi_{ijm}^{k\ell n}$, etc. (see Figure 9). In the following we restrict our computations for simplicity to the ground and singly excited configurations. The general solutions to the problem

$$\mathcal{H}\psi = E\psi$$

will accordingly have the form

$$\psi_G = {}^{(o)}A_G\Phi_G + \sum_{i}^{occ}\sum_{k}^{un} {}^{(o)}A_i^k\,\Phi_i^k + \ldots \quad = \psi_o \;;\quad E_o$$

$$\psi_n = {}^{(n)}A_G\Phi_G + \sum_{i}^{occ}\sum_{k}^{un} {}^{(n)}A_i^k\,\Phi_i^k + \ldots \quad = \psi_n \;;\quad \begin{cases} E_n \\ n = 1,2 \ldots \end{cases}$$

The designations "occupied" orbitals i and "unoccupied" orbitals k refer to the ground configuration (see Figure 9). The problem now consists in finding the eigenvalues E_n and expansion coefficients ${}^{(n)}A$ (n = 0,1,2 ...). This is done by diagonalizing the matrix of $\mathcal{H}$, i.e., in solving the secular equation, in the basis of the configurational functions Φ_G Φ_i^k, etc. (see also section III.1). This procedure is called <u>configuration interaction</u> and is a very general method for treating many-electron problems. If the configurational functions are in any way reasonably conceived, the coefficient ${}^{(o)}A_G$ should be large (see also section IV.1 and IV.3) and the coefficients ${}^{(n)}A_G$ should be small.

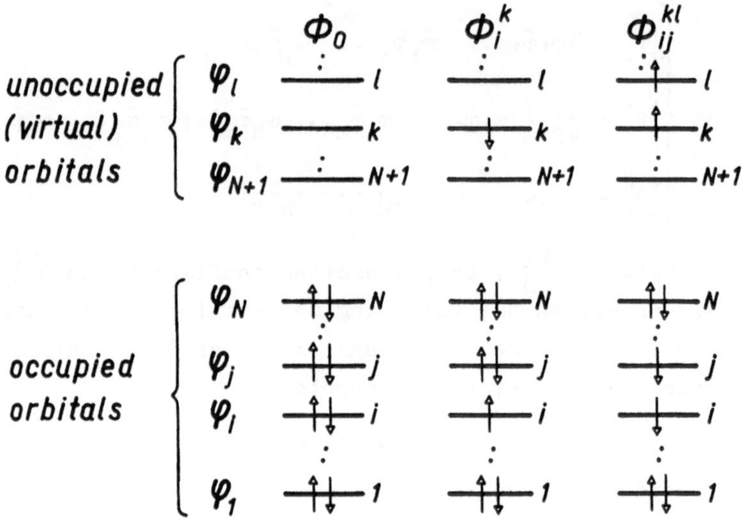

Figure 9

We now turn to the necessary evaluation of matrix elements: To this end we make use of the Slater-Condon rules for matrix elements between Slater determinants [6]. We find

$$
\begin{aligned}
\langle \Phi_G | \mathcal{H} | \Phi_G \rangle &= \sum_{i=1}^{N} 2\langle \varphi_i | h | \varphi_i \rangle \\
&+ \sum_{i=1}^{N} \sum_{j=1}^{N} \Big[2\langle \varphi_i(\mu)\varphi_j(\nu) | v | \varphi_i(\mu)\varphi_j(\nu) \rangle \\
&\qquad - \langle \varphi_i(\mu)\varphi_j(\nu) | v | \varphi_j(\mu)\varphi_i(\nu) \rangle \Big] \\
&\equiv \sum_i 2\langle i | h | i \rangle \\
&+ \sum_i \sum_j \left\{ 2\langle ij | ij \rangle - \langle ij | ji \rangle \right\} \\
&\equiv \sum_i 2h_{ii} + \sum_i \sum_j (2J_{ij} - K_{ij})
\end{aligned}
$$

Integrals of the type J_{ij} are called coulomb integrals, integrals of the type K_{ij} are designated exchange integrals. The double summations are here taken <u>independently</u> over <u>spatial</u> orbitals.

We now abbreviate

$$F_{ik} \equiv \langle i|h|k\rangle + \sum_{j=1}^{N} \left\{2\langle ij|kj\rangle - \langle ij|jk\rangle\right\} \quad \text{and find [7]:}$$

$$S = 0 \begin{cases} \langle \Phi_G|\mathcal{H}|{}^1\Phi_i^k\rangle = \sqrt{2}\, F_{ik} \\ \langle {}^1\Phi_i^k|\mathcal{H}|{}^1\Phi_i^k\rangle = \langle \Phi_G|\mathcal{H}|\Phi_G\rangle - F_{ii} + F_{kk} - \langle ik|ik\rangle + 2\langle ik|ki\rangle \\ \langle {}^1\Phi_i^k|\mathcal{H}|{}^1\Phi_m^k\rangle = -F_{mi} - \langle mk|ik\rangle + 2\langle mk|ki\rangle \\ \langle {}^1\Phi_i^k|\mathcal{H}|{}^1\Phi_i^n\rangle = F_{kn} - \langle ik|in\rangle + 2\langle ik|ni\rangle \\ \langle {}^1\Phi_i^k|\mathcal{H}|{}^1\Phi_m^n\rangle = - \langle mk|in\rangle + 2\langle mk|ni\rangle \end{cases}$$

$$S = 1 \begin{cases} \langle {}^3\Phi_i^k|\mathcal{H}|{}^3\Phi_i^k\rangle = \langle \Phi_G|\mathcal{H}|\Phi_G\rangle - F_{ii} + F_{kk} - \langle ik|ik\rangle \\ \langle {}^3\Phi_i^k|\mathcal{H}|{}^3\Phi_m^k\rangle = -F_{mi} - \langle mk|ik\rangle \\ \langle {}^3\Phi_i^k|\mathcal{H}|{}^3\Phi_i^n\rangle = F_{kn} - \langle ik|in\rangle \\ \langle {}^3\Phi_i^k|\mathcal{H}|{}^3\Phi_m^n\rangle = - \langle mk|in\rangle \end{cases}$$

These formulae are exact.

<u>Exercise</u> a) Verify the above expressions for the two-electron case and compare with section III.1.

b) Verify for the case of the 6-electron problem $|1\bar{1}\ 2\bar{2}\ 3\bar{3}|$.

3. The semiempirical PPP approximation for π electrons [8]

The matrix elements between configurational functions obtained in section III.2 lead to the evaluation of one- and two-electron integrals over one-electron functions (MO's). Here we consider the semiempirical evaluation of these integrals in the frame of π electron theory. Consequently we treat in a more general fashion some points already mentioned in section III.1., p. 23-26.

a) Two-electron integrals: We write

$$\langle \overset{(\mu)}{i} j | \underset{(\nu)}{k} \ell \rangle \equiv \int \varphi_i^*(\mu)\varphi_j^*(\nu) \frac{e^2}{r_{\mu\nu}} \varphi_k(\mu)\varphi_\ell(\nu) d\tau_\mu d\tau_\nu$$

and expand

$$\left.\begin{aligned} \varphi_i &= \sum_p c_{ip}\chi_p \\ \varphi_j &= \sum_q c_{jq}\chi_q \\ \varphi_k &= \sum_r c_{kr}\chi_r \\ \varphi_\ell &= \sum_s c_{\ell s}\chi_s \end{aligned}\right\} \quad \langle ij|k\ell\rangle = \sum_p\sum_q\sum_r\sum_s c_{ip}^* c_{jq}^* c_{kr} c_{\ell s} \langle pq|rs\rangle$$

$$\langle pq|rs\rangle \equiv \int \chi_p^*(1)\chi_q^*(2) \frac{e^2}{r_{12}} \chi_r(1)\chi_s(2) d\tau_1 d\tau_2$$

$\equiv$ (pr|qs) in Pariser-Parr notation [8]

The functions χ_p, χ_q here, of course, represent $2p_\pi$ AO's on atoms p,q respectively.

Neglect of differential overlap (ZDO-approximation):

$$\chi_p^*(1)\chi_r(1) = \delta_{pr} \cdot \chi_p^*(1)\chi_p(1)$$

leads to the simplification

$$\langle ij|k\ell\rangle = \sum_p\sum_q c_{ip}^* c_{jq}^* c_{kp} c_{\ell q} \langle pq|pq\rangle$$

$\langle pq|pq\rangle \equiv (pp|qq)$ is also often designated γ_{pq}.

An estimate of γ_{pp} (see also II.1, p. 24) is given by the relation

$$\gamma_{pp} \approx I_p - EA_p$$

where I_p, EA_p are the valence state ionization potential and electron affinity of atom p, respectively.

For γ_{pq} ($p \neq q$) Pariser and Parr suggest the "uniformly charged sphere" approximation (see Figure 10):

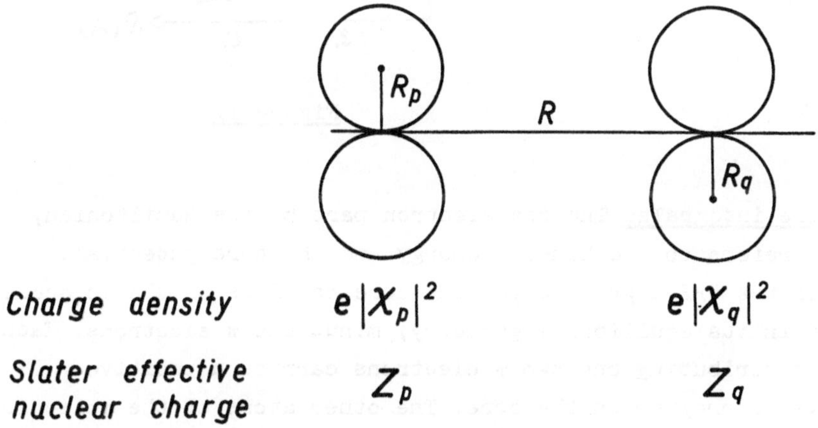

Figure 10

The radius R_p is given by

$$R_p = \frac{4{,}597}{Z_p} \times 10^{-8} \text{ cm}$$

The number 4.597 is obtained by setting the "electrostatic" value of γ_{pp} equal to the analytical value.

The simplified form of the electrostatic formula, valid for $R \geq 2.80$ Å reads:

$$\gamma_{pq} = \frac{7.1975}{R}\left\{\left[1+\left(\frac{R_p-R_q}{2R}\right)^2\right]^{-1/2} + \left[1+\left(\frac{R_p+R_q}{2R}\right)^2\right]^{-1/2}\right\} \text{ eV.}$$

(For $R < 2.80$ Å, see [8].)

Figure 11 shows the value of γ_{pq} where p and q refer to C atoms as a function of distance

$R = 0 : \gamma_{CC} \approx I_C - EA_C$

$R \to \infty : \gamma_{CC} \approx \frac{e^2}{R}$

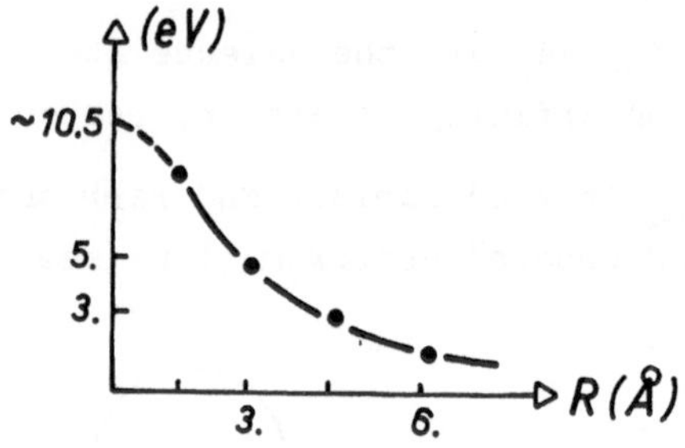

Figure 11

b) Core integrals: The one-electron part of the Hamiltonian, $h(\mu)$, refers to the kinetic energy and the core potential. Within the PPP approximation the core consists of the molecule, fixed in its equilibrium geometry, minus the π electrons. Each atom contributing one/two π electrons carries a positive charge of one/two in the core. The other atoms of the molecule (such as H atoms) also are part of the core, but are assumed neutral (see also III.1, page 25). Upon expansion

$$\langle i|h|k\rangle = \sum_p \sum_r c_{ip}^* c_{kr} \langle p|h|r\rangle$$

we now distinguish different cases

$$\langle p|h|r\rangle \equiv h_{pr} \begin{cases} p = r,\ h_{pp} \equiv \alpha_p \text{ ; see below} \\ p \text{ and } r \text{ are nearest neighbors, } h_{pr} = \beta_{pr} \\ p \text{ and } r \text{ are not n.n., } h_{pr} = 0 \end{cases}$$

β_{pr} is best calibrated on the spectroscopic properties of a test-molecule. It is generally of the order 2-3 eV.

$$\alpha_p \equiv \langle p|h_{core}|p\rangle$$

As already stated above and from the point of view of atom p we subdivide:

$$h_{core}(\mu) = T(\mu) + U_p(\mu) + \sum_{q\neq p} U_q(\mu) + \sum_r U'_r(\mu)$$

Kinetic term	Interaction with atomic core p	Interaction w. other atomic cores $q\neq p$	Interaction with neutral atoms attached to the core, such as H-atoms

As we have seen previously:

$$\{T(\mu) + U_p(\mu)\}\chi_p(\mu) = W_p \cdot \chi_p(\mu) , \qquad W_p = - I_p$$

I_p is the "valence state" ionization potential for an electron in the $2p_\pi$ orbital χ_p of the sp^2-hybridized atom p in the molecule.

Furthermore:

$$U_q(\mu) = U'_q(\mu) - z_q^{(\pi)} \int \chi_q^*(\nu)\chi_q(\nu) \frac{e^2}{r_{\mu\nu}} d\nu$$

i.e., the interaction with the core atom q is equal to the interaction with the neutral atom q, $U'_q(\mu)$, minus the interaction with the "missing" π electrons on that atom, the number of which being $z_q^{(\pi)}$. For carbon $z_q^{(\pi)} = 1$, for nitrogen $z_q^{(\pi)} = 1$ in pyridine, but $z_q^{(\pi)} = 2$ in pyrrhole, for instance.

Consequently:

$$\alpha_p = -I_p - \sum_{q\neq p} \{z_q^{(\pi)} \langle pq|pq\rangle - \langle p|U'_q|p\rangle\} + \sum_r \langle p|U'_r|p\rangle$$

Neglecting penetration integrals, i.e., integrals corresponding to the interaction with neutral atoms, or incorporating these terms implicitly into an effective valence state ionization potential I'_p, the formula simplifies to the "working" expression

$$\alpha_p = -I'_p - \sum_{q\neq p} z_q^{(\pi)} \cdot \gamma_{pq}$$

Let us illustrate this by some examples:

Benzene

$$\alpha_1 = -I_C^{(1)} - \gamma_{12} - \gamma_{13} - \gamma_{14} - \gamma_{15} - \gamma_{16}$$
$$= -I_C^{(1)} - 2\gamma_{12} - 2\gamma_{13} - \gamma_{14} ,$$

and $\alpha_1 = \alpha_2 = \ldots = \alpha_6$

Pyrrhole

$$\alpha_1 = -I_N^{(2)} - \gamma_{12} - \gamma_{13} - \gamma_{14} - \gamma_{15}$$
$$= -I_N^{(2)} - 2\gamma_{12} - 2\gamma_{13} , \text{ but}$$
$$\alpha_2 = -I_C^{(1)} - 2\gamma_{21} - \gamma_{23} - \gamma_{24} - \gamma_{25}$$

The nitrogen atom in pyrrhole contributes two π electrons, thus $I_N^{(2)}$ refers to a valence state ionization potential for double ionization. In the expression for α_2, γ_{21} must accordingly be counted twice, as $z_1^{(\pi)} = 2$.

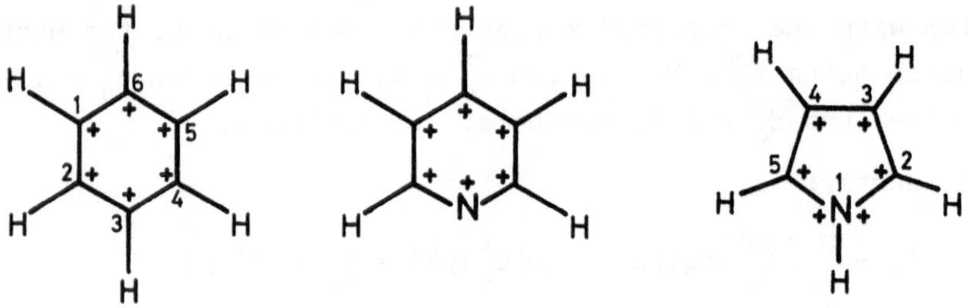

Figure 12, PPP cores

4. Benzene as an example

The ground configuration of the π electron system of benzene

$$\Phi_0 = |a\bar{a}e_{+1}\bar{e}_{+1}e_{-1}\bar{e}_{-1}|$$

is represented in Figure 13.

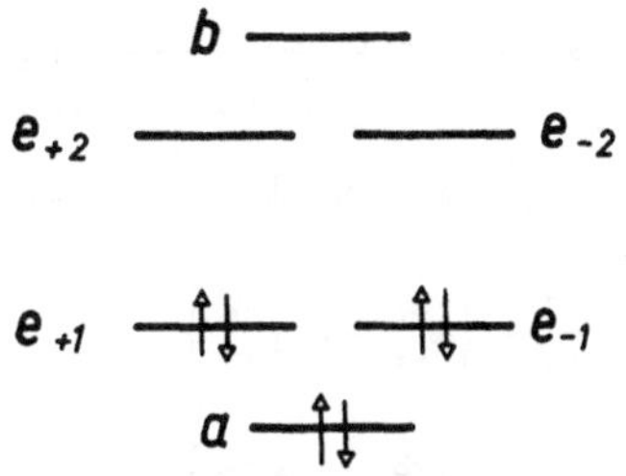

Figure 13

To discuss the lowest excited states we limit ourselves to considering only the four degenerate singly excited (singlet) configurations (see also Section II.2):

$$\Phi_{-1}^{-2} = \frac{1}{\sqrt{2}}\left\{|a\bar{a}e_{+1}\bar{e}_{+1}\overparen{e_{-1}\bar{e}_{-2}}| + |a\bar{a}e_{+1}\bar{e}_{+1}\overparen{e_{-2}\bar{e}_{-1}}|\right\}$$

$$\Phi_{+1}^{+2} = \frac{1}{\sqrt{2}}\left\{|a\bar{a}\overparen{e_{+1}\bar{e}_{+2}}e_{-1}\bar{e}_{-1}| + |a\bar{a}\overparen{e_{+2}\bar{e}_{+1}}e_{-1}\bar{e}_{-1}|\right\}$$

$$\Phi_{-1}^{+2} = \frac{1}{\sqrt{2}}\left\{|a\bar{a}e_{+1}\bar{e}_{+1}\overparen{e_{-1}\bar{e}_{+2}}| + |a\bar{a}e_{+1}\bar{e}_{+1}\overparen{e_{+2}\bar{e}_{-1}}|\right\}$$

$$\Phi_{+1}^{-2} = \frac{1}{\sqrt{2}}\left\{|a\bar{a}\overparen{e_{+1}\bar{e}_{-2}}e_{-1}\bar{e}_{-1}| + |a\bar{a}\overparen{e_{-2}\bar{e}_{+1}}e_{-1}\bar{e}_{-1}|\right\}$$

The one-electron functions (MO's) in their complex form are

$$e_j = \frac{1}{\sqrt{6}}\sum_p \omega^{jp}\chi_p \quad , \qquad j = 0, \pm 1, \pm 2, 3$$

and $\omega \equiv e^{2\pi i/6}$. Corresponding to their transformation properties under the point group C_6 we write $e_0 \equiv a$ and $e_3 \equiv b$.

Including the ground configuration, but neglecting higher configurations, the problem would lead to a 5×5 configuration interaction secular determinant. Full exploitation of symmetry simplifies the problem decisively, however. In the D_{6h} point symmetry of the molecule the ground configuration only mixes with doubly excited configurations and Φ_0 thus is a relatively acceptable description of the ground state. (See also Section IV.3 on the Brillouin theorem.)

$\psi_o \approx \Phi_o$, transforms like A_{1g} under D_{6h}. From the way the one-electron functions transform under D_{6h}, we may also deduce how the configurational functions transform [9]:

$$\psi_1 = \frac{1}{i\sqrt{2}}\left\{\Phi_{-1}^{+2} - \Phi_{+1}^{-2}\right\} \quad \text{transforms like } B_{2u}$$

$$\psi_2 = \frac{1}{\sqrt{2}}\left\{\Phi_{-1}^{+2} + \Phi_{+1}^{-2}\right\} \quad \text{transforms like } B_{1u}$$

$$\psi_3 = \frac{1}{i\sqrt{2}}\left\{\Phi_{-1}^{-2} - \Phi_{+1}^{+2}\right\}$$

$$\psi_4 = \frac{1}{\sqrt{2}}\left\{\Phi_{-1}^{-2} + \Phi_{+1}^{+2}\right\}$$

(ψ_3, ψ_4) transform like E_{1u} and must be degenerate

There are no matrix elements between functions transforming according to different irreducible representations of the same group. Consequently, the excitation energies are directly given by the diagonal matrix elements between the above functions. Taking this into account, and upon expansion:

$$E_1({}^1B_{2u}) = \langle\psi_1|\mathcal{H}^\pi|\psi_1\rangle = E_o + h_{22} - h_{11} + K_{o2} - K_{o3}$$

$$E_2({}^1B_{1u}) = \langle\psi_2|\mathcal{H}^\pi|\psi_2\rangle = E_o + h_{22} - h_{11} - K_{o2} + 3K_{o3}$$

$$E_3({}^1E_{1u}) = \langle\psi_{\frac{3}{4}}|\mathcal{H}^\pi|\psi_{\frac{3}{4}}\rangle = E_o + h_{22} - h_{11} + 2K_{o1} - K_{o3}$$

where, of course:

$$\mathcal{H}^\pi = \sum_{\mu=1}^{6} h(\mu) + \sum_{\mu>}^{6}\sum_{\nu=1}^{6} \frac{e^2}{r_{\mu\nu}}$$

Concerning the evaluation of the above matrix elements [5]

$$h_{ii} \equiv \langle e_i|h|e_i\rangle = \langle e_{-i}|h|e_{-i}\rangle .$$

A two-electron integral is written

$\langle e_i e_k|e_j e_\ell\rangle \equiv \langle ik|j\ell\rangle$, and with the ZDO approximation, is equal to the expansion

$$\langle ik|j\ell\rangle = \frac{1}{36}\sum_p\sum_q \omega^{(j-i)p+(\ell-k)q} \cdot \gamma_{pq}$$

It may be shown that due to the cyclic symmetry this integral vanishes, unless

$(j-i) = (k-\ell)$, or $(j-i) = (k-\ell) \pm 6n$, where $n = 1,2,\ldots$

All Coulomb integrals $\langle ik|ik\rangle$ are equal, and the exchange integrals $\langle ik|ki\rangle$ reduce to three types, namely K_{01}, K_{02} and K_{03}, where, for instance:

$$\left.\begin{array}{ll} \overset{i\;k}{\langle 0\underline{+}1} | \overset{j\;\ell}{\underline{+}1\;0\rangle} & = K_{01} \\ \langle \underline{+}1\underline{+}2 | \underline{+}2\underline{+}1\rangle & = K_{01} \end{array}\right\} = \frac{1}{6}\left\{\gamma_{11} + \gamma_{12} - \gamma_{13} - \gamma_{14}\right\}$$

and less obviously:

$$\langle -2-1|+2+1\rangle \text{ modulo } 6 = K_{02} = \frac{1}{6}\left\{\gamma_{11} - \gamma_{12} - \gamma_{13} + \gamma_{14}\right\}$$

$$\langle -2-1|+1+2\rangle \text{ modulo } 6 = K_{03} = \frac{1}{6}\left\{\gamma_{11} - 2\gamma_{12} + 2\gamma_{13} - \gamma_{14}\right\}$$

In conclusion we obtain the following energy level diagram (Figure 14).

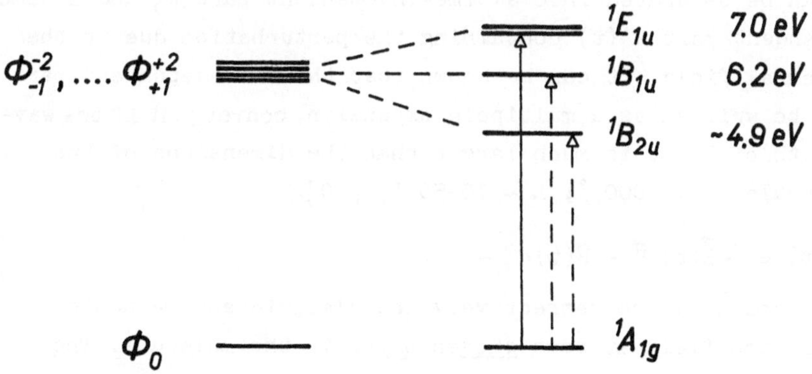

Figure 14

Listed are the experimental energies above the ground state. For a comment on the transition probabilities, see the next section.

<u>Exercise:</u> Verify the energy expression for the $^1B_{2u}$ state.

5. Electric-dipole transition probability

The semiclassical Hamiltonian for a many-electron system in a radiation field (within the B.O.-approximation) is written [2]

$$\mathcal{H} = \sum_{\mu} \frac{1}{2m} \left\{ \vec{p}_\mu - \frac{e}{c} \vec{A}_\mu(t) \right\}^2 + V$$

where $\vec{A}_\mu(t)$ is the vector potential of the field at the site of electron μ at time t. V contains all electrostatic terms. e designates the (negative) charge of the electron.

$$V = \sum_{\mu} U(\mu) + \sum_{\mu > \nu}\sum \frac{e^2}{r_{\mu\nu}}$$

$\mathcal{H}$ can be separated into a time-independent part $\mathcal{H}_0$ and a time-dependent part $\mathcal{H}'(t)$ containing the perturbation due to the external field. It can be shown that the time-dependend part may be written as a multipole expansion, convergent if the wavelength of light is much larger than the dimensions of the molecule ($\lambda > 1000$ Å; $L \sim 10$-50 Å) [10]:

$$\mathcal{H}'(t) = -\vec{E}(t)\cdot\vec{R} - \vec{H}(t)\cdot\vec{M} + \ldots.$$

$\vec{E}(t)$ and $\vec{H}(t)$ are respectively the electric and magnetic radiation field at some <u>chosen point</u> in the molecule. The probability for a transition from a (ground) state ψ_a to an excited state ψ_b is by time-dependent perturbation theory [2] proportional to:

$$W_{a\to b} \sim |\langle\psi_a|\mathcal{H}'(t)|\psi_b\rangle|^2 .$$

In this expression the contribution of the electric dipole term turns out to be by far the most important one. We consequently write, after averaging over randomly oriented molecules:

$$W_{a\to b} \sim |\langle\psi_a|\vec{R}|\psi_b\rangle|^2 \equiv D_{ab}$$

D_{ab} is called the dipole strength and $\vec{R}$ is the electric dipole operator: $\vec{R} \equiv \sum_{\mu} e\vec{r}_{\mu}$.

The integral $\langle\psi_a|\vec{R}|\psi_b\rangle$ is called the <u>transition moment</u>.

For a many-electron (2N) system we find:

$$\psi_a \approx \Phi_o = |\varphi_1\bar{\varphi}_1\ldots\varphi_i\bar{\varphi}_i\ldots\varphi_N\bar{\varphi}_N|$$

and

$$\psi_b \approx B_o\Phi_o + \sum_i\sum_k B_i^k\Phi_i^k + \ldots$$

where as usual

$$\Phi_i^k = \frac{1}{\sqrt{2}}\left\{|\varphi_1\bar{\varphi}_1\ldots\varphi_i\bar{\varphi}_k\ldots\varphi_N\bar{\varphi}_N| + |\varphi_1\bar{\varphi}_1\ldots\varphi_k\bar{\varphi}_i\ldots\varphi_N\bar{\varphi}_N|\right\} .$$

Consequently

$$\langle\psi_a|\vec{R}|\psi_b\rangle = B_o\langle\Phi_o|\vec{R}|\Phi_o\rangle + \sum_i\sum_k B_i^k\langle\Phi_o|\vec{R}|\Phi_i^k\rangle .$$

As B_o is in general small, the first term (where $\langle\Phi_o|\vec{R}|\Phi_o\rangle$ within our approximations is the expression for the dipole moment of the ground state) may be neglected.

$$\langle\Phi_o|\vec{R}|\Phi_i^k\rangle = \sqrt{2}\,e\langle\varphi_i|\vec{r}|\varphi_k\rangle$$

Thus to a good approximation

$$\langle \psi_a | \vec{R} | \psi_b \rangle = \sqrt{2}\, e \sum_i \sum_k B_i^k \langle \varphi_i | \vec{r} | \varphi_k \rangle$$

The problem then boils down to the evaluation of matrix elements of the operator $\vec{r}$ between one-electron MO's.

$$\langle \varphi_i | \vec{r} | \varphi_k \rangle = \sum_p \sum_q c_{ip}^* \, c_{kq} \, \langle \chi_p | \vec{r} | \chi_q \rangle$$

In the frame of PPP calculations on π electrons the following procedure is admissible to estimate orders of magnitude:

$$\langle \chi_p | \vec{r} | \chi_p \rangle = \vec{r}_{op}$$

$\vec{r}_{op}$ is the position vector of atom p with respect to the origin of the coordinate system. This expression is exact. The next expression is only exact between like atoms and orbitals.

$$\langle \chi_p | \vec{r} | \chi_q \rangle \approx S_{pq} \left(\frac{\vec{r}_{op} + \vec{r}_{oq}}{2} \right)$$

If one is schematically consistent in neglecting differential overlap, these cross-terms may even be neglected.

Some <u>symmetry considerations</u>:

A transition a→b is called electric dipole allowed if the integral $\langle \psi_a | \vec{R} | \psi_b \rangle \equiv \int \psi_a^* \, \vec{R} \, \psi_b \, d\tau$ fails to vanish. Not to vanish, this integral must transform according to the totally symmetric irreducible representation of the point group of the molecule. On the other hand, the integral $\langle \psi_a | \vec{R} | \psi_b \rangle$ transforms like the triple direct product of the irreducible representations according to which ψ_a, $\vec{R}$ and ψ_b transform respectively. Thus $\Gamma_a \otimes \Gamma_{\vec{R}} \otimes \Gamma_b$ must contain the totally symmetric irreducible representation.

To revert to benzene as an example: $\Gamma_a = A_{1g}$, so $\Gamma_{\vec{R}} \otimes \Gamma_b$ must contain A_{1g} for the triple product to contain A_{1g}. This can only be the case if $\Gamma_{\vec{R}} = \Gamma_b$. The electric dipole operator $\vec{R}$ transforms like the vector components x,y,z. The components x,y in the plane of the molecule transform like E_{1u}. So Γ_b must be identical with E_{1u} for a transition to be electric dipole allowed in the plane of the molecule. We then summarize

Transition $A_{1g} \rightarrow E_{1u}$	electric dipole allowed in the plane of the molecule
Transitions $A_{1g} \rightarrow B_{2u}$, $A_{1g} \rightarrow B_{1u}$	electric dipole forbidden

If the latter transitions still appear in the spectrum, this is due to <u>vibronic coupling</u> with normal modes of appropriate symmetry.

IV. Self-consistent-field (SCF)-methods

SCF calculations essentially follow the method proposed by Hartree and Fock about 40 years ago and applied by these authors to atoms. It is a variational procedure (analogous to the simple Ritz method, see section II.1., page 5) taking to some extent electron interaction into account explicitly [11].

1. Simple LCAO-formulation of the closed-shell case

We consider a 2N electron system and assume that it can be described by an antisymmetrized function

$$\Phi_o = |\varphi_1\bar{\varphi}_1 \cdots\cdots \varphi_i\bar{\varphi}_i \cdots\cdots \varphi_N\bar{\varphi}_N|$$

The Hamiltonian be:

$$\mathcal{H} = \sum_{\mu=1}^{2N} h(\mu) + \sum_{\mu>\nu}^{2N} \frac{e^2}{r_{\mu\nu}}$$

We now seek a set of one electron functions $\varphi_1 \ldots \varphi_i \ldots \varphi_N$ such that

$$E_o = \frac{\int \Phi_o^* \mathcal{H} \Phi_o d\tau}{\int \Phi_o^* \Phi_o d\tau} \qquad \text{be a minimum.}$$

From the variational principle we may assume that

$$E_{o\ min} \longrightarrow E_G \quad ;$$

the minimized energy will approximate the true ground state energy.

We must, of course, impose the constraint that the set of φ_i be orthonormal, that is

$$\langle\varphi_i|\varphi_j\rangle = \delta_{ij} \ , \quad \text{for all } \ i,j$$

Assuming Φ_o to be normalized, we find (as on page 28):

$$E_o = \langle \Phi_o | \mathcal{H} | \Phi_o \rangle =$$

$$= \sum_{i=1}^{N} 2\langle \varphi_i | h | \varphi_i \rangle + \sum_{i=1}^{N} \sum_{j=1}^{N} \{ \underbrace{2\langle \varphi_i \varphi_j | v | \varphi_i \varphi_j \rangle}_{\text{Coulomb-term}} - \underbrace{\langle \varphi_i \varphi_j | v | \varphi_j \varphi_i \rangle}_{\text{Exchange-term}} \}$$

The double summations go independently over spatial orbitals. With the LCAO-expansion:

$$\varphi_i = \sum_{p=1}^{M} c_{ip} \chi_p \qquad \text{M: number of basis functions}$$

$$\varphi_j = \sum_{q=1}^{M} c_{jq} \chi_q$$

$$\varphi_i^* = \sum_{r}^{M} c_{ir}^* \chi_r^*$$

$$\varphi_j^* = \sum_{s}^{M} c_{js}^* \chi_s^*$$

We may write:

$$\langle \varphi_i | h | \varphi_i \rangle = \sum_r \sum_p c_{ir}^* c_{ip} h_{rp}$$

$$\langle \varphi_i \varphi_j | \varphi_i \varphi_j \rangle = \sum_r \sum_s \sum_p \sum_q c_{ir}^* c_{js}^* c_{ip} c_{jq} \gamma_{rspq}$$

$$\langle \varphi_i \varphi_j | \varphi_j \varphi_i \rangle = \sum_r \sum_s \sum_q \sum_p c_{ir}^* c_{js}^* c_{jq} c_{ip} \gamma_{rsqp}$$

where $h_{rp} \equiv \langle \chi_r | h | \chi_p \rangle$

$$\gamma_{rspq} \equiv \langle \overset{(1)}{\chi_r} \overset{(2)}{\chi_s} | \overset{(1)}{\chi_p} \overset{(2)}{\chi_q} \rangle \; ; \qquad \gamma_{rsqp} \equiv \langle \overset{(1)}{\chi_r} \overset{(2)}{\chi_s} | \overset{(1)}{\chi_q} \overset{(2)}{\chi_p} \rangle$$

Introducing these expressions into the one for E_o, we see that E_o becomes a function of the coefficients:

$E_o\ (c_{ip};\ i=1...N,\ p=1...M)$ or equivalently

$E_o\ (c^*_{ir};\ i=1...N,\ r=1...M)$.

To minimize this function under the constraints

$$f_{ij} \equiv \sum_r \sum_q c^*_{ir} c_{jq} S_{rq} = \delta_{ij}$$

we obtain the equation below.

$$\frac{\partial E_o}{\partial c^*_{ir}} - \underbrace{\sum_{j=1}^{N} \lambda_{ij} \frac{\partial f_{ij}}{\partial c^*_{ir}}}_{\text{takes into account N orthonormality constraints}} = 0 \quad ; \quad r = 1 \ldots\ldots M$$

λ_{ij} is a Lagrangian multiplier. It may be shown (see section IV.2) that one may set

$$\lambda_{ij} = \delta_{ij} \cdot 2\varepsilon_i$$

thereby simplifying our equation to

$$\frac{\partial E_o}{\partial c^*_{ir}} - 2\varepsilon_i \frac{\partial f_{ii}}{\partial c^*_{ir}} = 0$$

For computational reasons we assume that we consider a particular value of the indices i and r.

To remember that we consider particular values of i and r let us set $i \equiv i'$, $r \equiv r'$ and evaluate the above expression. In differentiating the different terms of E_o we must exercise some caution. For instance, in the exchange part

$$\sum_{ij} \sum_{rsqp} c^*_{ir} c^*_{js} c_{jq} c_{ip} \gamma_{rsqp}$$

the term in $c^*_{i'r'}$ will occur twice, once for $i=i'$, $r=r'$, and once for $j=i'$, $s=r'$, leading to the respective derivatives

$$\sum_{j}\sum_{sqp} c^*_{js}c_{jq}c_{i'p}\gamma_{r'sqp} \qquad \text{and}$$

$$\sum_{i}\sum_{rqp} c^*_{ir}c_{i'q}c_{ip}\gamma_{rr'qp} .$$

It is easily seen that both terms are equal, leading thus to a factor of 2 in the general expression. A similar situation is of course encountered in the coulomb part. We thus obtain:

$$2\sum_{p} c_{i'p}h_{r'p} + 4\sum_{p} c_{i'p}\sum_{j}\sum_{sq} c^*_{js}c_{jq}\gamma_{r'spq}$$

$$- 2\sum_{p} c_{i'p}\sum_{j}\sum_{sq} c^*_{js}c_{jq}\gamma_{r'sqp}$$

$$- 2\epsilon_{i'}\sum_{p} c_{i'p}S_{r'p} = 0$$

Dividing by two and abbreviating

$$\sum_{j=1}^{N} c^*_{js}c_{jq} \equiv D_{sq} , \qquad \text{we find:}$$

$$\left(\equiv \frac{1}{2} P_{sq}\right)$$

$$\sum_{p} c_{i'p}\left\{h_{r'p} + 2\sum_{sq} D_{sq}\gamma_{r'spq} - \sum_{sq} D_{sq}\gamma_{r'sqp} - S_{r'p}\epsilon_{i'}\right\} = 0$$

The index i' is now redundant and $\left\{\begin{matrix} p = 1\ldots M \\ r'=r=1\ldots M\end{matrix}\right\}$. We thus have M linear homogeneous equations with M unknowns. The existence of nontrivial solutions requires the determinant of the coefficients to vanish:

$$\det\left|\underbrace{h_{rp} + 2\sum_{sq} D_{sq}\gamma_{rspq} - \sum_{sq} D_{sq}\gamma_{rsqp}}_{F_{rp}} - S_{rp}\epsilon\right| = 0$$

$p = 1 \ldots M$, $r = 1 \ldots M$. The first three terms are customarily designated by F_{rp} and are the matrix element of the Fock-operator F in the basis χ_r, χ_p. To solve the problem one must:

1. Choose an appropriate basis χ_p. The bigger M, the better. M should always be significantly greater than N. For M=N no energy lowering will be achieved. In this case the energy will remain constant, as $\langle \Phi_0 | \mathcal{H} | \Phi_0 \rangle$ remains invariant under a unitary transformation of the $\varphi_1 \ldots \varphi_N$ among themselves.
2. Compute integrals γ_{rspq}, γ_{rsqp}, h_{rp}, S_{rp}. However, the factors D_{sq} couple the equations together and require a knowledge of the solutions $\varphi_i = \sum_p c_{ip}\chi_p$. One therefore proceeds as follows:
3. Guess an approximate form for the φ_i and compute approximate D_{sq}.
4. Solve the secular problem a first time.
5. Recompute the D_{sq} with the new eigenvectors.
6. Solve the secular problem a second time.
7. Repeat the procedure until eigenvalues ϵ and eigenvectors converge (generally about 20 times). Then self-consistency has been attained.

One thus obtains M SCF-eigenvalues: $\epsilon_1 \ldots \epsilon_M$, corresponding each to a respective one-electron SCF-function or SCF-MO: $\varphi_1^{SCF}, \varphi_2^{SCF} \ldots \varphi_M^{SCF}$. Of these MO's the N lowest ones are doubly filled to approximate the many-electron ground state. The M-N higher ones are virtual orbitals.

If the number of basis functions M becomes quasi-infinite, one approaches the true SCF energy or Hartree-Fock limit. By judicious choice of the basis functions one often succeeds in ab initio calculations to come close to the Hartree-Fock limit, even if M is finite.

2. Semiempirical simplification (ZDO-approximation)

Setting in all integrals $\chi_p^*(\mu)\cdot\chi_q(\mu) = \delta_{pq}\chi_p^*(\mu)\cdot\chi_p(\mu)$, the γ_{rspq} are neglected except for r=p, s=q, and the γ_{rsqp} are neglected except for q=r, s=p. Remembering that for a particular matrix element r and p are fixed indices, s and q running indices, the eigenvalue equation simplifies to:

$$\det\left| h_{rp} + 2\delta_{rp}\sum_s D_{ss}\gamma_{rsrs} - D_{pr}\gamma_{rprp} - \delta_{rp}\varepsilon \right| = 0$$

Abbreviating $\gamma_{rsrs} \equiv \gamma_{rs}$, $\gamma_{rprp} \equiv \gamma_{rp}$, one obtains:

$$\left\{\begin{array}{ll} F_{pp} = h_{pp} + 2\sum_s D_{ss}\gamma_{ps} - D_{pp}\gamma_{pp} , & \\ F_{rp} = h_{rp} - D_{pr}\gamma_{rp} , & r \neq p \end{array}\right\}$$

D_{pr} is here defined as $\sum_j c_{jp}^* c_{jr}$

For real orbitals $D_{pr}=D_{rp}$.

3. More general formulation of the closed-shell case

In the previous sections we optimized the one-electron functions by varying LCAO-coefficients solely. While this is very often done in practice, the Hartree-Fock problem may, however, be more generally formulated without specifying how the φ_i are varied to minimize E_o. We start out from our energy expression:

$$E = E_o = \sum_{i=1}^{N} 2\langle\varphi_i|h|\varphi_i\rangle + \sum_{i=1}^{N}\sum_{j=1}^{N}\left\{2\langle\varphi_i\varphi_j|v|\varphi_i\varphi_j\rangle - \langle\varphi_i\varphi_j|v|\varphi_j\varphi_i\rangle\right\}$$

and consider the functions φ_i, φ_j^* as independent variables. To minimize this expression with respect to these variables under imposition of the orthonormality constraints $f_{ij} \equiv \langle\varphi_i|\varphi_j\rangle = \delta_{ij}$, we may set the following total differential equal to zero:

$$\sum_i \frac{\partial E}{\partial\varphi_i}\delta\varphi_i + \sum_i \frac{\partial E}{\partial\varphi_i^*}\delta\varphi_i^* - \sum_i\sum_j \frac{\partial f_{ij}}{\partial\varphi_i}\lambda_{ij}\delta\varphi_i - \sum_i\sum_j \frac{\partial f_{ij}}{\partial\varphi_i^*}\lambda_{ij}^*\delta\varphi_i^* = 0$$

This is fulfilled if the factor of any $\delta\varphi_i$ identically vanishes. We thus obtain a series of equations:

$$\frac{\partial E}{\partial \varphi_i^*} - \sum_j \frac{\partial f_{ij}}{\partial \varphi_i^*} \lambda_{ji} = 0 \,, \qquad i = 1 \ldots N \,,$$

which leads to:

(To verify this following expression, write out E and f_{ij} not in terms of brackets, but of integrals. Perform the differentiation within the integrals and set the sum of integrands equal to zero.)

$$\left\{2h + 4\sum_j \langle\varphi_j|v|\varphi_j\rangle\right\} \varphi_i - 2\sum_j \langle\varphi_j|v|\varphi_i\rangle \varphi_j - \sum_j \varphi_j \lambda_{ji} = 0$$

We also obtain a set of equivalent complex conjugate equations. Dividing by 2, setting $\frac{1}{2}\lambda_{ji} \equiv \epsilon_{ji}$ and abbreviating

$$\langle\varphi_j|v|\varphi_j\rangle \varphi_i \equiv J_j\varphi_i \quad , \qquad \langle\varphi_j|v|\varphi_i\rangle \varphi_j \equiv K_j\varphi_i$$

We use the Coulomb operator J_j and exchange operator K_j:

$$\left\{h + \sum_j (2J_j - K_j)\right\} \varphi_i = \sum_j \epsilon_{ji}\varphi_j \,.$$

It may now be shown

a) That the matrix of the ϵ_{ji} is Hermitian and may be brought into diagonal form by a unitary transformation of the φ_i among themselves.

b) That such a unitary transformation leaves the Fock operator operating on a function φ

$$\left\{h + \sum_j (2J_j - K_j)\right\} \varphi \equiv F\varphi \qquad \text{invariant.}$$

We thus may assume a priori the φ_i to be in the proper form to write:

$$\left\{h + \sum_j (2J_j - K_j)\right\} \varphi_i \equiv F\varphi_i = \epsilon_i\varphi_i \,.$$

These pseudo-eigenvalue equations for the φ_i are fulfilled if the φ_i are self-consistent. If not, we must solve the equation $F\varphi$ iteratively until the above relation is fulfilled (see previous section).

From above it follows immediately that

$$\langle\varphi_i|F|\varphi_i\rangle = \epsilon_i, \quad \text{and} \quad \langle\varphi_i|F|\varphi_j\rangle = \epsilon_i\langle\varphi_i|\varphi_j\rangle = 0 \quad (\text{for } i \neq j)$$

This relation of course also holds for matrix elements of F between filled and eventual virtual orbitals. We now previously found (section III.2, page 29):

$$\langle\Phi_G|\mathcal{H}|\Phi_i^k\rangle = \sqrt{2}\, F_{ik} \equiv \sqrt{2}\,\langle\varphi_i|F|\varphi_k\rangle .$$

It immediately follows that if the φ_i, φ_k are obtained by the same SCF calculation, then these matrix elements vanish. This situation is summarized as <u>Brillouin's theorem</u>: Matrix elements between a closed-shell SCF ground state and <u>singly</u> excited configurations defined within the same set of SCF orbitals <u>vanish</u>.

(This theorem may also be stated in the reverse way. For instance, we notice that the symmetry orbitals of benzene are SCF orbitals within the π electron approximation, because all matrix elements between the ground and singly excited configurations vanish.)

From

$$\sum_i \langle\varphi_i|h + \sum_j (2J_j - K_j)|\varphi_i\rangle \equiv \sum_i \langle\varphi_i|F|\varphi_i\rangle = \sum_i \epsilon_i$$

and as

nota bene ↙

$$E_o^{SCF} = \sum_{i=1}^{N} \langle\varphi_i|2h + \sum_{j=1}^{N} (2J_j - K_j)|\varphi_i\rangle ,$$

we immediately find:

$$E_o^{SCF} = \sum_{i=1}^{N}\Big(\langle i|F|i\rangle + \langle i|h|i\rangle\Big) = \sum_{i=1}^{N}(\epsilon_i+h_{ii}) \ .$$

(This contrasts with the simple Hückel-type one-electron approximation, where we had

$$E_o^{Hückel} = \sum_{i=1}^{N} 2\epsilon_i \ .)$$

The true SCF ground state energy is of course not the true energy, as the interaction with doubly excited and other configurations is left out. One generally defines within the Born-Oppenheimer approximation

Exact nonrelativistic energy minus Hartree-Fock energy = Correlation energy

4. Koopmans' "theorem"

Monoionization from a neutral closed-shell molecule may, to an acceptable degree of approximation (with notable and important exceptions), be pictured as the extraction of an electron out of a given SCF orbital, the other electrons remaining unaffected [11].

Suppose that the electron comes from a SCF orbital i (see Figure 16), the energy of the ion will in that approximation be given by (upper index: 1 singlet, 2 doublet, K Koopmans)

$$^2E_{ion}^{K} = {}^1E_o^{SCF} - \Big[\langle i|h|i\rangle + \sum_{j=1}^{N}\Big\{2\langle ij|ij\rangle - \langle ij|ji\rangle\Big\}\Big]$$

Within this approximation it follows immediately that

$$^2E_{ion}^{K} = {}^1E_o^{SCF} - \epsilon_i \quad , \text{ or}$$

$$\epsilon_i = {}^1E_o^{SCF} - {}^2E_{ion}^{K} = -I_k \ , \qquad k + i = N + 1$$

I_k being the ionization potential. Consequently

$(-\epsilon_N)$ gives an approximation of the ionization potential to the ground state of the monopositive ion, I_1.

$(-\epsilon_{N-1})$ { = I_2, approximates the ionization potential to the first excited state of the monopositive ion,

$(-\epsilon_{N-2})$ { = I_3, approximates the ionization potential to the second excited state, etc.

Koopmans' approximation neglects (Figure 17):

a) the reorganization energy of the electrons in the ion. This reorganization energy is taken into account in the restricted open-shell SCF energy of the ion (see Section IV.6.1).

b) the difference between the electron correlation energy of the neutral molecule and of the ion.

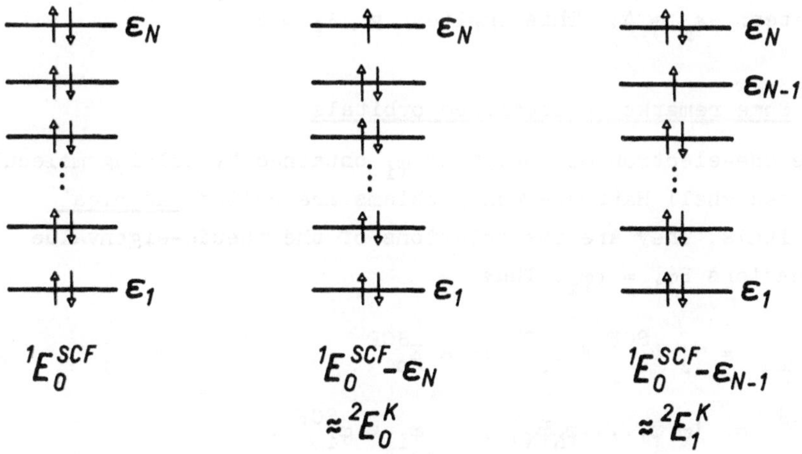

Figure 16 Schematic representation of the Koopmans' energy of a monopositive ion $^2E^K$.

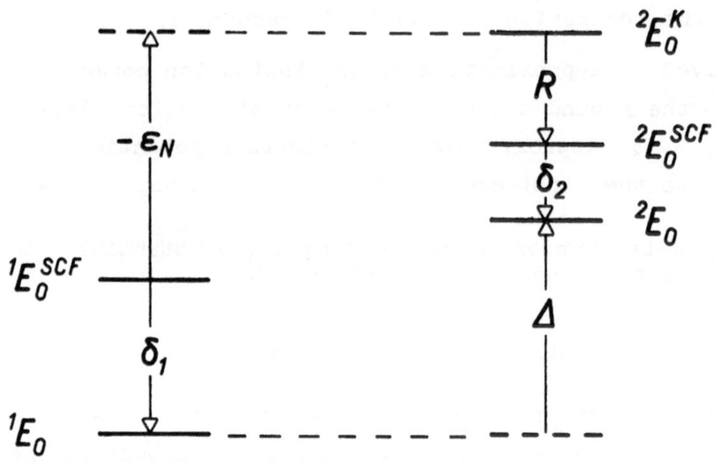

Figure 17 δ_1: Correlation energy of the neutral closed-shell molecule. R: Reorganization energy of the ground state of the monopositive ion. δ_2: Correlation energy of the ground state of the monopositive ion. Δ: Exact ionization energy to the ground state of the monopositive ion. Koopmans' "theorem" states: $-\epsilon_N \approx \Delta$. This implies: $R + \delta_2 \approx \delta_1$.

5. Some remarks on localized orbitals

The one-electron SCF orbitals φ_i obtained by solving molecular closed-shell Hartree-Fock-problems are called canonical orbitals. They are the solutions of the pseudo-eigenvalue equations $F\varphi_i = \epsilon\varphi_i$. Thus

$$E_{o\,min} = \int \Phi_o^{SCF}\, \mathcal{H}\, \Phi_o^{SCF}\, d\tau = E_o^{SCF}$$

$$\Phi_o^{SCF} = |\varphi_1\bar{\varphi}_1 \cdots \varphi_N\bar{\varphi}_N| \, , \qquad \varphi_i \equiv \varphi_i^{SCF} \, .$$

Now it may be shown that Φ_o^{SCF} can be expressed in terms of any set of orbitals φ_i' obtained by a unitary transformation of the (doubly) filled canonical φ_i^{SCF} among themselves, to yield the same many-electron energy E_o^{SCF} [11]. These new orbitals φ_i' of course no longer satisfy the pseudo-one-electron equations.

It has for instance proven instructive to construct non-canonical orbitals according to certain physical criteria, such as the criterion of maximization of electrostatic orbital self-energy D, where [12]

$$D = \sum_{i=1}^{N} \langle \varphi_i' \varphi_i' | v | \varphi_i' \varphi_i' \rangle$$

This corresponds to a minimization of interorbital electron interactions. Orbitals φ_i' so chosen can be expected to show nearly minimum interorbital correlation effects. Such orbitals turn out in general to be relatively strongly localized in certain parts of the molecule and are therefore called localized orbitals. Their interest lies in the possible transferability of localized orbitals between different molecules and the eventual transferability of correlation corrections.

There are, of course, other transformation or localization procedures than the one mentioned, and in considering localized orbitals it is therefore important to always enquire about the localization criterion.

6. Open-shell SCF methods

6.1. The restricted open-shell SCF method

Suppose we are interested in computing the lowest triplet state of a molecule. This may either be done by pure CI or it may be approached by optimizing one-electron functions in a manner analogous to the closed-shell HF-method. We thus need an open-shell SCF procedure. If we require our many-electron function from the start to correspond to a definite spin-state, our open shell procedure will, from that point of view, be called restricted. The restricted open-shell SCF procedure in its most widely applied form is mainly due to Roothaan [13].

We write our triplet functions as:

$$ {}^{3}_{-1}\Phi^{n}_{m} = |\varphi_1\bar{\varphi}_1\cdots\cdot\varphi_g\bar{\varphi}_g\bar{\varphi}_m\bar{\varphi}_n| $$

$$ {}^{3}_{0}\Phi^{n}_{m} = \frac{1}{\sqrt{2}}\left\{|\varphi_1\bar{\varphi}_1\cdots\cdot\varphi_g\bar{\varphi}_g\varphi_m\bar{\varphi}_n| - |\varphi_1\bar{\varphi}_1\cdots\cdot\varphi_g\bar{\varphi}_g\varphi_n\bar{\varphi}_m|\right\} $$

$$ {}^{3}_{+1}\Phi^{n}_{m} = |\varphi_1\bar{\varphi}_1\cdots\cdot\varphi_g\bar{\varphi}_g\varphi_m\varphi_n| $$

corresponding to a situation as shown in Figure 18.

We want to minimize

${}^{3}E = \langle {}^{3}\Phi|\mathcal{H}|{}^{3}\Phi\rangle$, subject to the constraints $\langle\varphi_i|\varphi_j\rangle = \delta_{ij}$

We find for the energy

$$ {}^{3}E = \Big[2\sum_{k=1}^{g} h_{kk} + \sum_{k=1}^{g}\sum_{\ell=1}^{g}\{2J_{k\ell}-K_{k\ell}\} \qquad \longleftarrow \text{closed shell terms} $$

$$ + h_{mm} + h_{nn} + J_{mn} - K_{mn} \qquad \longleftarrow \text{open shell terms} $$

$$ + \sum_{k=1}^{g}\{2J_{km}-K_{km}\} + \sum_{k=1}^{g}\{2J_{kn}-K_{kn}\}\Big] \longleftarrow \left\{\begin{array}{l}\text{closed-open shell}\\ \text{coupling terms}\end{array}\right. $$

The index g designates the highest doubly filled orbital, the indices m and n the two singly filled orbitals; the running indices k and ℓ run over the doubly filled (closed shell) spatial orbitals only; the indices i and j run over all spatial orbitals.

The procedure to find a pseudo-one-electron Fock operator for such an open-shell situation is analogous to the closed-shell case, with the added complication that it proves difficult to get rid of the nondiagonal Lagrangian multiplyers which couple closed and open-shell orbitals. By some clever but somewhat tedious algebraic manipulations [13] this may be achieved, leading, for our particular triplet case, to the open-shell Fock operator of the form: (OS ≡ open shell)

$$F^{OS} = h + 2J_T - K_T + 2M_T - 2K_O \quad , \quad \text{where}$$

$$J_T = \sum_{k=1}^{g} J_k + \frac{1}{2}(J_m+J_n)$$

$$K_T = \sum_{k=1}^{g} K_k + \frac{1}{2}(K_m+K_n)$$

$$K_O = \frac{1}{2}(K_m+K_n)$$

$$M_T = \sum_{k=1}^{g} M_k + \frac{1}{2}(M_m+M_n) \quad , \quad \text{and where}$$

$$M_i\varphi = \frac{1}{2}\left\{\langle\varphi_i|K_m+K_n|\varphi\rangle\,\varphi_i + \langle\varphi_i|\varphi\rangle\,(K_m+K_n)\,\varphi_i\right\}$$

We then have the pseudo-eigenvalue equations

$$F^{OS}\varphi = \varepsilon^{OS}\cdot\varphi$$

The iterative method of solution (for instance within the frame of an LCAO expansion) is similar to the closed-shell case. The corresponding eigensolutions then of course satisfy the relations

$$F^{OS}\varphi_i = \varepsilon_i\varphi_i \qquad ; \qquad \langle\varphi_i|F|\varphi_j\rangle = \varepsilon_i\delta_{ij}$$

The g lowest φ_i correspond to optimized doubly filled orbitals and φ_m (where m=g+1) and φ_n (where n=g+2) to the optimized singly occupied ones. The energy $\langle{}^3\Phi|\mathcal{H}|{}^3\Phi\rangle$ computed from these orbitals has attained a relative minimum.

In the closed-shell SCF case, by Brillouin's theorem, we have vanishing matrix elements between the ground state (i.e. ground configuration) and singly excited configurations defined within the same set of SCF MO's.

Such a general Brillouin theorem does not hold for open shells. In our triplet SCF case one can prove that matrix elements vanish between the minimized triplet ${}^3\Phi_m^n$ function and other triplet functions with the same number of singly occupied orbitals, one of which must be either m or n: ${}^3\Phi_m^p$, ${}^3\Phi_\ell^n$, ${}^3\Phi_{mm'}^{np}$, ${}^3\Phi_{m\ell}^{nn}$. Figure 18 illustrates this in the case of the six-electron problem. See also ref. [18].

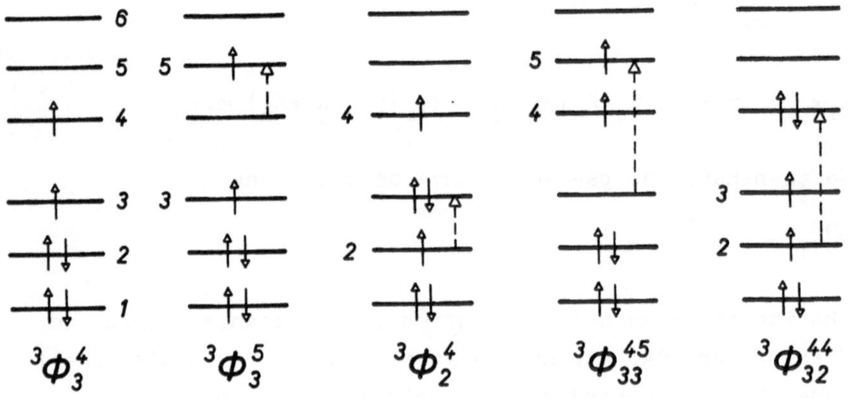

Figure 18

Exercise

1. Consider the triplet SCF state ${}^3\Phi_3^4$, $|1\bar{1}\ 2\bar{2}\ 3\ 4|$. Prove that $\langle {}^3\Phi_3^4|\mathcal{H}|{}^3\Phi_3^5\rangle = \langle 4|F^{OS}|5\rangle = 0$, where ${}^3\Phi_3^5$ is defined in terms of the SCF orbitals of ${}^3\Phi_3^4$.

In the case of a doublet we have only one singly occupied orbital, and the SCF equations are accordingly simpler than in the case of the triplet.

(In reference to Roothaan's paper [13]: To obtain F^{OS} for the doublet case requires the same parameters as for the triplet, namely $f = \frac{1}{2}$ $a = 1$, $b = 2$; $\alpha = 0$, $\beta = -2$.)

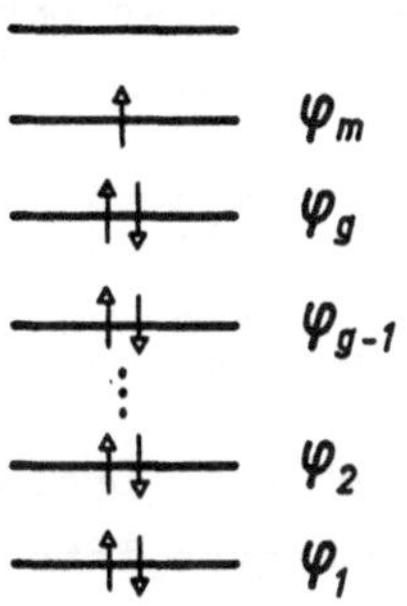

Figure 19

The difference between the Koopmans' energy of the ion (see Section IV.4) and the restricted open-shell SCF energy is called the reorganization energy.

6.2. The unrestricted open-shell SCF method

In this method the energy of an arbitrary system of M electrons with α spin and N electrons with β spin is minimized [14].

The many-electron function Φ is written

$\Phi = \left|\varphi_1\varphi_2\ldots\varphi_M\bar{\varphi}_{M+1}\bar{\varphi}_{M+2}\ldots\bar{\varphi}_{M+N}\right|$, where the normalization factor of the Slater determinant of course is $\left\{(M+N)!\right\}^{-1/2}$.
The total energy may be wr

$$E = \langle\Phi|\mathcal{H}|\Phi\rangle$$

$$= \sum_i^{\alpha+\beta} h_{ii} + \frac{1}{2}\sum_i^{\alpha+\beta}\sum_j^{\alpha+\beta} J_{ij} - \frac{1}{2}\left(\sum_i^{\alpha}\sum_j K_{ij} + \sum_i^{\beta}\sum_j^{\beta} K_{ij}\right)$$

where $\sum_i^{\alpha+\beta}$ means summation over all spin-orbitals with both α and β spin

$\sum_i^{\alpha}$ means summation over all spin-orbitals with only α spin, etc.

In the double summation the indices take on values independently of each other.

Minimization of $\langle\Phi|\mathcal{H}|\Phi\rangle$ leads to separate Fock equations for orbitals with α and with β spin

$$F^{\alpha}\varphi_i^{\alpha} = \epsilon_i^{\alpha}\varphi_i^{\alpha}$$

$$F^{\beta}\varphi_k^{\beta} = \epsilon_k^{\beta}\varphi_k^{\beta}$$

This is accordingly also called the method of "different orbitals for different spins".

Although this procedure may take electron correlation to a higher degree into account than the restricted one, Φ is not an eigenfunction of S^2 and therefore in itself not a physically acceptable solution. Once Φ has been optimized, eigenfunctions of S^2 must be projected out of it.

V. All-valence MO procedures

As discussed in the introductory chapter, we subdivide the molecular electrons into groups. From an energy criterion, we in general merely distinguish between atomic core electrons (for instance, 1s electrons of second row atoms) and valence electrons (2s, 2p electrons of second row atoms, 1s electrons of hydrogen). A further subdivision, into σ and π electrons for instance, is of course only possible in the presence of an appropriate element of symmetry. The atomic nuclei and the core electrons are assumed to be frozen into a static core. The B.O.Hamiltonian of the valence electrons reads:

$$\mathcal{H}_{Val} = T_{Val} + V_{Val\text{-}Core} + V_{Val\text{-}Val}$$

$$= \sum_{\mu}^{(Val)} h_{Core}(\mu) + \sum_{\mu>\nu}^{(Val)}\sum \frac{e^2}{r_{\mu\nu}}$$

1. The Extended Hückel (EH) Method

This is the all-valence electron analogue of the ordinary Hückel method. Accordingly, one defines an effective one-electron Hamiltonian

$$\mathcal{H}_{\substack{Val\\eff}} = \sum_{\mu}^{(Val)} h_{eff}(\mu)$$

The first comprehensive application of this procedure to organic molecules is due to Hoffmann [15], but similar schemes were followed previously by Wolfsberg and Helmholz [16] and others to inorganic molecules.

For simple hydrocarbons diagonal atomic matrix elements $h_{qq} \equiv \langle \chi_q | h_{eff} | \chi_q \rangle$ are generally set equal to valence state ionization potentials

Hydrogen	$\langle 1s\|h\|1s \rangle$	= -13.6 eV
Carbon	$\langle 2s\|h\|2s \rangle$	= -21.4 eV
	$\langle 2p\|h\|2p \rangle$	= -11.4 eV

while for nondiagonal elements h_{qr} a variety of modifications occur in the literature [15]:

$$h_{qr} = k \cdot S_{qr}(h_{qq}+h_{rr}) \cdot \frac{1}{2} \qquad \text{I}$$

$$h_{qr} = -k \cdot S_{qr}(h_{qq} \cdot h_{rr})^{1/2} \qquad \text{II}$$

$$h_{qr} = k \cdot S_{qr} \cdot 2 \frac{h_{qq} \cdot h_{rr}}{(h_{qq}+h_{rr})} \qquad \text{III}$$

$$h_{qr} = (k-|S_{qr}|) \cdot S_{qr}(h_{qq}+h_{rr}) \cdot \frac{1}{2} \qquad \text{IV}$$

In case IV the off-diagonal elements do not automatically insure invariance of the eigenvalues with respect to a rotation of the coordinate axis of reference. k is an adjustable parameter and is taken to be 1.75 in case I and between 1.7 and 2.5 in the other cases. As in ordinary Hückel theory, the problem boils down to solving the eigenvalue equation

$$\det|h_{qr} - S_{qr}\epsilon| = 0$$

The overlap integrals S_{qr} are calculated exactly from Slater orbitals, with an exponent $\mu_C = 1.625$ for carbon and μ_H between 1.00 and 1.20 for hydrogen. In π electron theory we have only to consider π-type overlap (see Figure 20), whereas in the present case we have all possible combinations of π and σ-type overlap. The neglect of overlap leads here to meaningless results. The total electronic energy is just the sum of one-electron energies: $E = \sum_i b_i \epsilon_i$, b_i being the occupation number. For molecules with heteroatoms certain iterative variants of the EH method have been tested to improve charge distributions. The diagonal matrix elements are modified by the atomic charges on the respective atoms until a (limited) self-consistency is achieved.

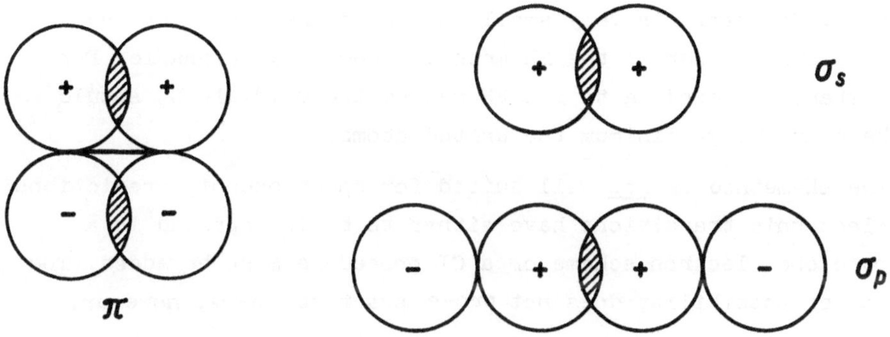

Figure 20 π and σ-type overlap

The EH method has proven very successful in predicting most stable molecular conformations and in studying the local and overall symmetry of molecular orbitals, in particular in showing the extent of delocalization of σ orbitals. The success in predicting relative conformational energies is difficult to explain in a precise way.

To change the conformation of the nuclei implies (within the B.O. approximation) not only changing the total electronic energy, but also the purely classical inter-nuclear (or inter-core) respulsion. That is, we have to include the term $\sum_{A>B}\sum \frac{z_A z_B e^2}{R_{AB}}$ (see page 1) in our energy change, where z_A and z_B are atomic core charge numbers. We thus consider the overall Hamiltonian:

$$\mathcal{H}_{overall} = \mathcal{H}_{Val} + V_{\substack{Core-\\Core}} = \sum_{\mu}^{(Val)} h(\mu) + \sum_{\mu>\nu}^{(Val)}\sum \frac{e^2}{r_{\mu\nu}} + \sum_{A>B}\sum \frac{z_A z_B e^2}{R_{AB}}$$

It appears a posteriori that with a change of conformation the effective one-electron Hamiltonian, or rather the total energy computed from it, varies more like $\mathcal{H}_{overall}$ than as $\mathcal{H}_{Val}$. However, for very small molecules the conformational predictive power of the EH method becomes questionable. For instance, according to the EH method the molecule H_2 should have an energy minimum for united atoms.

The EH-method is not well suited for spectroscopic predictions: Electronic transitions have either to be interpreted in a pure one-electron scheme, or a CI procedure must be added. The latter possibility does not prove practical here, however.

2. Electronic population analysis [17]

The electronic population analysis gives a systematic procedure for the interpretation of LCAO-MO data. Consider a simple MO, such as

$$\varphi = c_q \chi_{qA} + c_r \chi_{rB} \quad ; \quad \begin{cases} \chi_q \text{ being on atom A} \\ \chi_r \text{ being on atom B} \end{cases}$$

and suppose the MO is occupied ny N (≡ b of sections II.3 and V.1; we here use the notation of Mulliken) electrons. We then have (all functions are assumed real):

$$N\varphi^2 = Nc_q^2(\chi_{qA})^2 + 2Nc_q c_r \chi_{qA}\chi_{rB} + Nc_r^2(\chi_{rB})^2$$

Integrating each term over all space gives

$$N = Nc_q^2 + 2Nc_q c_r S_{qr} + Nc_r^2$$

Net atomic population on atom A	overlap population	Net atomic population on atom B

We also define the gross atomic population on atom

A as $N(c_q^2 + c_q c_r S_{qr})$ and on atom

B as $N(c_r^2 + c_q c_r S_{qr})$.

We now generalize for the case of an MO φ_i of a polyatomic molecule, to which every atom contributes more than one AO. $\left(\sum_q\right)$ implies summation over all orbitals on a given atom, $\left(\sum_A\right)$ implies summation over all atoms. Thus

$$\varphi_i = \sum_q \sum_A c_{i,qA} \chi_{qA} , \qquad \text{and consequently}$$

$$N_i = N_i \left\{ \sum_q \sum_A c_{i,qA}^2 + \sum_q \sum_r \sum \sum_{A \neq B} c_{i,qA} c_{i,rB} S_{qA,rB} \right\}$$

There are no terms between different orbitals on the same atom because the corresponding overlap integrals vanish. Based on the above expression we define:

Net atomic population on atom A for orbital i:

$$N_i \sum_q c_{i,qA}^2$$

Gross atomic population on atom A for orbital i:

$$N_i \left\{ \sum_q c_{i,qA}^2 + \sum_q \sum_r \sum_{B(\neq A)} c_{i,qA} c_{i,rB} S_{qA,rB} \right\}$$

The sum of this latter expression over all atoms is just N_i. Summation over the filled levels i of the net and gross atomic populations then gives the total net atomic and total gross atomic populations on atom A respectively.

3. Semiempirical all-valence calculations, including electron interaction

In the all-valence approach a molecule like ethylene,-which in section III.1 we considered as a two-electron problem-, has twelve electrons. Considering an AO basis consisting of the $2s, 2p_x, 2p_y, 2p_z$ orbitals of the two carbon atoms and the 1s orbitals of the four hydrogen atoms, our SCF matrix will have the dimension 12 x 12. However, we have seen in the expression for F_{rp} (see page 45) that many-electron integrals have to be evaluated over all combinations of basis orbitals which,-in spite of the fact that many integrals will be equal-, implies the order of 12^4 integrals. By invoking the ZDO approximation as defined on page 47, this number is reduced by a factor of 12^2. This exemplifies the fact that to keep calculations on relatively large molecules tractable, approximations of the ZDO-type may be important.

The following sections are devoted to short descriptions of such simplified computational procedures.

3.1. The CNDO (complete neglect of differential overlap) method

The approximations are [18]:

1) The χ_p are treated as if they form an orthonormal set; thus S_{pq} is set equal to δ_{pq}.

2) All two-electron integrals which depend on the overlapping of charge densities of different basis orbitals are neglected. This means that

$$\langle pq|rs\rangle = \delta_{pr}\delta_{qs}\langle pq|pq\rangle \equiv \delta_{pr}\delta_{qs}\gamma_{pq}$$

3) The electron interaction integrals γ_{pq} are assumed to depend only on the atoms to which the orbitals χ_p and χ_q belong and not to the actual type of orbital. Thus γ_{pq} is set equal to γ_{AB}, measuring an average repulsion between

an electron in a valence atomic orbital on A and another in a valence orbital on B. (The justification for this approximation will be given below.)

4) The core matrix element h_{pp} contains the interaction energy of an electron in valence orbital χ_p on A with the core of A and with the cores of all other atoms B. It may be written

$$h_{pp} = \langle p | -\frac{\hbar^2}{2m}\nabla^2 - V_A | p \rangle - \sum_{B(\neq A)} \langle p | V_B | p \rangle$$ and is simplified to

$$h_{pp} = U_{pp} - \sum_{B(\neq A)} V_{AB}$$

as $\langle p | V_B | p \rangle$ is considered to be the same for all valence atomic orbitals on A.

U_{pp} is essentially an atomic quantity, measuring the energy of an electron in the atomic orbital χ_p on the core of A. (See p.33: Note the difference with the PPP method in defining the core.)

5) Core matrix elements h_{pq}, where χ_p and χ_q are different but both belong to A, may in analogy to 4) be written:

$$h_{pq} = U_{pq} - \sum_{B(\neq A)} \langle p | V_B | q \rangle$$

However, due to the mutual orthogonality of s, p_x, p_y, p_z, U_{pq} is exactly equal to zero, and the remaining terms are small, so that one sets

$$h_{pq} = 0 \quad \text{for } p \neq q, \quad \chi_p \text{ and } \chi_q \text{ on A.}$$

6) Core matrix elements h_{pr}, where χ_p is on A and χ_r is on B will for simplicity be considered proportional to the overlap integral S_{pr}:

$$h_{pr} = \beta^o(A,B,R_{AB})\, S_{pr}$$

$\beta^o(A,B,R_{AB})$ is a parameter dependent on the nature of atoms A and B, and eventually on their separation, but not on the form of orbitals χ_p and χ_r.

With these approximations the Fock matrix elements take on the following form:

$$F_{pp} = U_{pp} - \sum_{B(\neq A)} V_{AB} + 2\sum_{s}^{(A)} D_{ss}\gamma_{AA} + 2\sum_{s'}^{(B)} D_{s's'}\gamma_{AB} - D_{pp}\gamma_{AA}$$

$$= U_{pp} - \sum_{B(\neq A)} V_{AB} + 2\, D_{AA}\gamma_{AA} + 2\sum_{B(\neq A)} D_{BB}\gamma_{AB} - D_{pp}\gamma_{AA}$$

or, writing $2\, D_{AA} \equiv P_{AA}$, $2\, D_{BB} \equiv P_{BB}$ (see also page 45):

$$F_{pp} = U_{pp} + \left(P_{AA} - \frac{1}{2} P_{pp}\right) \gamma_{AA} + \sum_{B(\neq A)} \left(P_{BB}\gamma_{AB} - V_{AB}\right)$$

$$F_{pq} = \beta^{o}(A,B)S_{pq} - \frac{1}{2} P_{pq}\gamma_{AB} \qquad \chi_p,\chi_q \text{ real}$$

The expression for F_{pq} ($p\neq q$) applies even if p and q are on the same atom. Then $S_{pq} = 0$, and γ_{AB} is replaced by γ_{AA}.

Parametrization of quantities:

Pople and Segal [18] calculate for second-row atoms:

γ_{AA} exactly as $\langle 2s_A\, 2s_A | 2s_A\, 2s_A \rangle$

γ_{AB} exactly as $\langle 2s_A\, 2s_B | 2s_A\, 2s_B \rangle$

U_{pp} from valence-state ionization potentials, using the computed value of γ_{AA}.

V_{AB} as $\langle s_A(1) \mid \frac{z_B e^2}{r_{1B}} \mid s_A(1) \rangle$, where z_B is the core charge number of B

$\beta^{o}(A,B) = \frac{1}{2}\,(\beta^{o}_{A} + \beta^{o}_{B})$, where β^{o}_{A} and β^{o}_{B} are atomic parameters calibrated on small molecules.

- Other approximations for the γ:

  $$\gamma_{AB} = \gamma^{th} - \alpha \exp\,(-\eta\, R^{n}_{AB})$$

  γ^{th} is the corresponding exact (analytic) quantity. α, η, and n are adjustable parameters. For $R_{AB} = 0$, $\gamma = \gamma^{th} - \alpha$. The parameter α may be interpreted as a local "correlation term".

- Other approximation for V_{AB}:

  $$V_{AB} = z_B \cdot \gamma_{AB} + P$$

  P refers to the interaction of an electron on A with the neutral atom B. It is thus a penetration term (and is eventually neglected). Note the similarity to the PPP approach (see Section III.3, p. 33).

4. Invariance of the (exact) SCF eigenvalue problem to unitary basis transformations

We consider a set of closed-shell SCF orbitals φ_i, defined in a basis χ_p. In deviation from our usual convention we write

$$\varphi_i = \sum_p \chi_p\, c_{pi}$$

inverting the indices of the coefficients c. In matrix form

$$(\varphi) = (\chi)(C)$$

(φ) is the row vector of the φ_i ($i = 1 \ldots M$), (χ) the row vector of the χ_p, and (C) the coefficient matrix of the c_{pi}. This enables us to write the Hartree-Fock equations in a convenient matrix form:

$$(F)(C) = (S)(C)(\varepsilon)$$

(F) is the matrix F_{rp} of the Fock operator in the basis of the χ_p, (S) the overlap matrix S_{rp}, (ϵ) the diagonal eigenvalue matrix. If $\{c_i\}$ represents the column vector of the coefficients of φ_i, then $(F)\{c_i\} = (S)\{c_i\}\epsilon_i$, where ϵ_i is the (scalar) ith eigenvalue.

We now express our basis (χ) in terms of a new basis (χ'):

$$\chi_p = \sum_\alpha \chi'_\alpha u_{\alpha p} \quad ; \quad (\chi) = (\chi')(U)$$

Thus $(\varphi) = (\chi')(U)(C) = (\chi')(C')$

The matrices F and S are consequently also to be referred to the new basis. We find (see below):

$$(F) = (U)^+(F')(U) \quad , \quad (S) = (U^+)(S')(U)$$

where $(U)^+$ is the conjugate transposed (or adjoint) of (U), and (F') and (S') are defined in the basis (χ').

For simplicity, we assume both (χ) and (χ') to form an orthonormal basis. Thus (C) then is unitary, and (S) becomes the unit matrix. (C') must also be unitary and consequently also (U).

Thus: $(U)^+ = (U)^{-1}$

and we obtain

$$(U)^{-1}(F')(U)(C) = (C)(\epsilon)$$

Multiplying both sides by (U) and remembering that (U)(C) = (C'), we obtain

$$(F')(C') = (C')(\epsilon) \qquad \text{or equivalently}$$

$$(F')\{c'_i\} = \{c'_i\} \cdot \epsilon_i$$

The SCF equations are now defined in the basis (χ') and must lead to the <u>same</u> eigenvalues as the ones defined in the basis (χ).

It now remains to be shown that (F') with respect to (χ') has the <u>same form</u> as (F) with respect to (χ) (see p. 45):

$$F_{rp} = \Big[\langle \chi_r|h|\chi_p\rangle + \sum_s\sum_q\sum_j c^*_{sj}\, c_{qj}\ \cdot$$

$$\cdot\ \big\{2\langle \chi_r\chi_s|\chi_p\chi_q\rangle - \langle \chi_r\chi_s|\chi_q\chi_p\rangle\big\}\Big]$$

With $\chi_p = \sum_\alpha \chi'_\alpha u_{\alpha p}$, $\chi_q = \sum_\beta \chi'_\beta u_{\beta q}$

$$\chi^*_r = \sum_\gamma \chi'^*_\gamma u^*_{\gamma r}\ , \qquad \chi^*_s = \sum_\delta \chi'^*_\delta u^*_{\delta s}$$

and noting that by definition

$$\sum_s u^*_{\delta s}\, c^*_{sj} = c'^*_{\delta j}\ , \qquad \sum_q u_{\beta q}\, c_{qj} = c'_{\beta j}$$

and $\sum_{j=1}^{N} c'^*_{\delta j}\, c'_{\beta j} \equiv D'_{\delta\beta}\left(= \frac{1}{2} P'_{\delta\beta}\right)$

we obtain

$$F_{rp} = \sum_\gamma\sum_\alpha u^*_{\gamma r}\, u_{\alpha p}\Big[\langle \chi'_\gamma|h|\chi'_\alpha\rangle +$$

$$+ \sum_\delta\sum_\beta D'_{\delta\beta}\big\{2\langle \chi'_\gamma\chi'_\delta|\chi'_\alpha\chi'_\beta\rangle - \langle \chi'_\gamma\chi'_\delta|\chi'_\beta\chi'_\alpha\rangle\big\}\Big]$$

Thus $F_{rp} = \sum_\gamma\sum_\alpha u^*_{\gamma r}\, u_{\alpha p}\, F'_{\gamma\alpha}$, or

$$F = (U)^+(F')(U)$$

which completes our demonstration.

In case all the coefficients are real, the word "unitary" in this section may be replaced by "orthogonal", the word "adjoint" by the word "transposed".

Invariance of integrals in the CNDO approximation:

If approximation 2) of the CNDO scheme is applied without further conditions, the integrals do not necessarily transform to preserve the invariance of the Fock equations with respect to unitary transformations of the basis.

Consider the integral $\langle p_{xA}\, s_B | p_{yA}\, s_B \rangle$ which, according to approximation 2), is neglected. Consider the following unitary basis transformation: Rotate the (local) coordinate system by 45° clockwise around the z axis.

p_{xA} goes into $\frac{1}{\sqrt{2}}\left(p'_{xA} + p'_{yA}\right)$, p_{yA} into $\frac{1}{\sqrt{2}}\left(- p'_{xA} + p'_{yA}\right)$

and the corresponding integral goes into

$$- \tfrac{1}{2} \langle p'_{xA}\, s'_B | p'_{xA}\, s'_B \rangle + \tfrac{1}{2} \langle p'_{xA}\, s'_B | p'_{yA}\, s'_B \rangle$$

$$- \tfrac{1}{2} \langle p'_{yA}\, s'_B | p'_{xA}\, s'_B \rangle + \tfrac{1}{2} \langle p'_{yA}\, s'_B | p'_{yA}\, s'_B \rangle$$

of which the first and last terms are not neglected and do not necessarily cancel. The necessary invariance is, however, restored by adopting an even cruder approximation, i.e. by setting

$$\langle p'_{xA}\, s'_B | p'_{xA}\, s'_B \rangle = \langle p'_{yA}\, s'_B | p'_{yA}\, s'_B \rangle \equiv \gamma_{AB} .$$

Now the transformed integral becomes zero through cancellation.

In this general way we ensure a pseudo-unitary transformation of the Fock operator, and it may be shown that the SCF eigenvalues remain invariant under the transformation considered.

VI. Some special topics

1. Optical activity

A medium is called optically active if the index of refraction (n) for left (ℓ) circularly polarized light is different from that for right (r) circularly polarized light:

$$\Delta n = n_\ell - n_r \neq 0$$

The measurement of this difference as a function of wavelength λ, $\Delta n(\lambda)$, is called optical rotatory dispersion (ORD). Directly connected to this effect is the fact that in regions of absorption the extinction coefficient (ϵ) for left and right circularly polarized light will also differ:

$$\Delta\epsilon(\lambda) = \epsilon_\ell(\lambda) - \epsilon_r(\lambda)$$

This latter phenomenon is called circular dichroism (CD). Inside an absorption band ORD will be anomalous, that is, there will be an inversion of sign. The combined effect of CD and anomalous ORD inside a region of absorption is called a Cotton effect.

Optical activity is a molecular effect. A molecule is optically active when it cannot be superimposed onto its mirror image. Such a molecule may not have a rotation-reflection axis S_n ($S_1 \equiv \sigma$, $S_2 \equiv i$). Many molecules occurring in living organisms are optically active.

Every transition $a \rightarrow b_i$ in an optically active molecule makes a certain contribution to $\Delta\epsilon$ and Δn. A CD/ORD spectrum where these contributions are clearly resolved may appear as shown in Figure 21. The transition $a \rightarrow b_1$ leads to a positive Cotton effect, the transition $a \rightarrow b_2$ to a negative one. While CD may effectively only be measured in regions of absorption, ORD curves have long tails outside of regions of absorption which are the superposition of the contributions of different transitions.

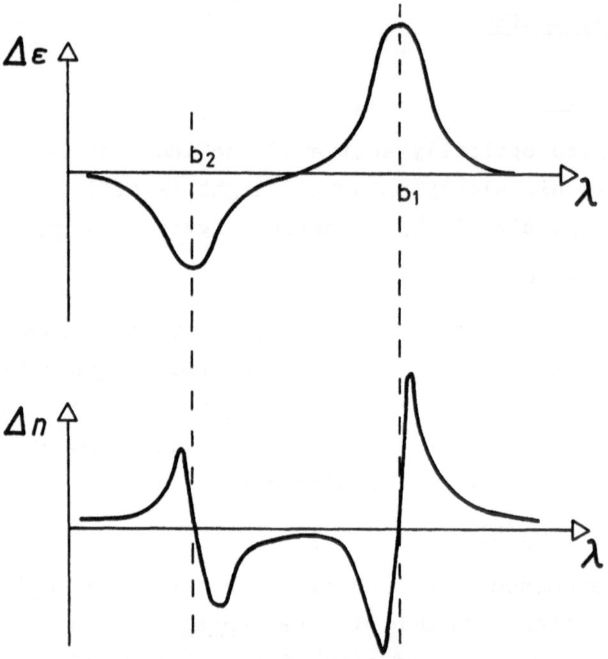

Figure 21

Quantum mechanically it may be shown [2,19] that the contribution which a given transition makes to the CD/ORD spectrum is proportional to a quantity called the rotatory strength $\mathbb{R}_{ab}$:

$$\mathbb{R}_{ab} \equiv \mathrm{Im}\left\{\langle\psi_a|\vec{R}|\psi_b\rangle\langle\psi_b|\vec{M}|\psi_a\rangle\right\}$$

$\vec{R}$ is the electric dipole operator (see Section III.5.) and $\vec{M}$ the magnetic dipole operator:

$$\vec{M} = \sum_\mu \vec{m}_\mu = \sum_\mu \frac{e}{2mc}\,\vec{\ell}_\mu$$

For simplicity we assume here the summation to go only over all electrons; we neglect vibronic effects due to the nuclei. e and m stand for charge and mass of the electron, $\vec{\ell}_\mu \equiv -i\hbar\, \vec{r}_\mu \times \vec{\nabla}_\mu$ is the angular momentum operator of the μth electron.

Im $\{ \}$ means that the imaginary part of the quantity in brackets is taken. The rotatory strength is actually a second-rank tensor, but for a system composed of many identical randomly oriented molecules one may consider the trace of this tensor. It is a pseudo-scalar, being the scalar product of a polar (electric dipole transition moment) and of an axial vector (magnetic dipole transition moment).

The connection between the rotatory strength and the experimentally determined quantity $\Delta\varepsilon(\lambda)$ is given by the proportionality (~)

$$\mathbb{R}_{ab} \sim \int_{\text{Band}} \frac{\Delta\varepsilon(\lambda)}{\lambda}\, d\lambda$$

For CD and ORD the rotatory strength plays a role formally comparable to the one of the dipole strength for ordinary absorption and dispersion (see Section III.5.):

$$D_{ab} \equiv \text{Re}\left\{\langle\psi_a|\vec{R}|\psi_b\rangle\langle\psi_b|\vec{R}|\psi_a\rangle\right\} \sim \int_{\text{Band}} \frac{\varepsilon(\lambda)}{\lambda}\, d\lambda$$

Because of the different transformation properties of $\vec{R}$ and $\vec{M}$ under S_n it may be proven that the rotatory strength always vanishes for systems containing such symmetry elements.

We now wish to show that even with extremely crude wavefunctions but which correctly reflect local symmetry properties, a semiquantitative discussion of optical activity is possible.

<u>Case 1:</u> The optical activity of the carbonyl $n \rightarrow \pi^*$ transition in a ketone (aldehyde). The one-electron energy level scheme of interest is depicted at left in Figure 22. We assume the corresponding many-electron states to be well represented by single-configuration functions. We also consider the highest filled (π and n) and lowest unfilled (π^*) MO's to be markedly

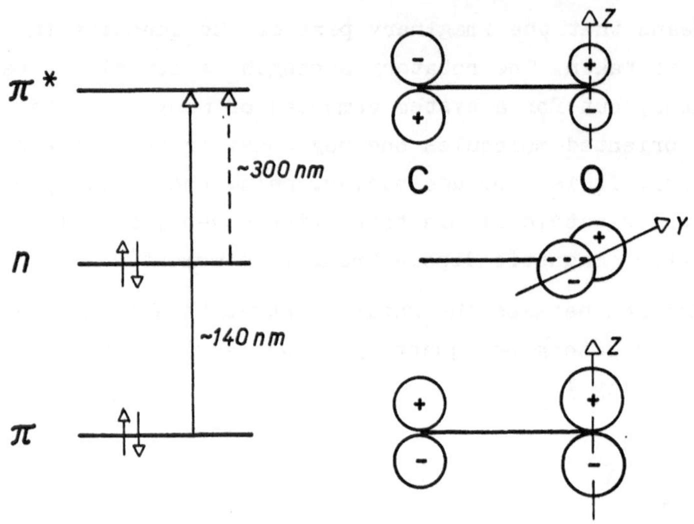

Figure 22

localized on the carbonyl chromophore, as depicted at right in Figure 22. In a symmetric ketone, of symmetry C_{2v}, for instance, one finds for the transition moments

$$\langle n|\vec{r}|\pi^*\rangle = 0\ , \quad \langle n|\vec{m}|\pi^*\rangle \neq 0\ , \qquad \mathbb{R}_{n\pi^*} = 0$$

$$\langle \pi|\vec{r}|\pi^*\rangle \neq 0\ , \quad \langle \pi|\vec{m}|\pi^*\rangle = 0\ , \qquad \mathbb{R}_{\pi\pi^*} = 0$$

Thus the $n \rightarrow \pi^*$ transition, occurring experimentally at $\lambda \approx 300$ nm, is magnetic dipole allowed and electric dipole forbidden, while for the $\pi - \pi^*$ transition, occurring at much shorter wavelength, it is the opposite. The rotatory strength vanishes in both cases.

Now suppose that we perturb the carbonyl group by introducing a substituent reducing the overall symmetry to C_1 (Figure 23a). The perturbing substituent will have the effect of slightly mixing some π character to the n orbital:

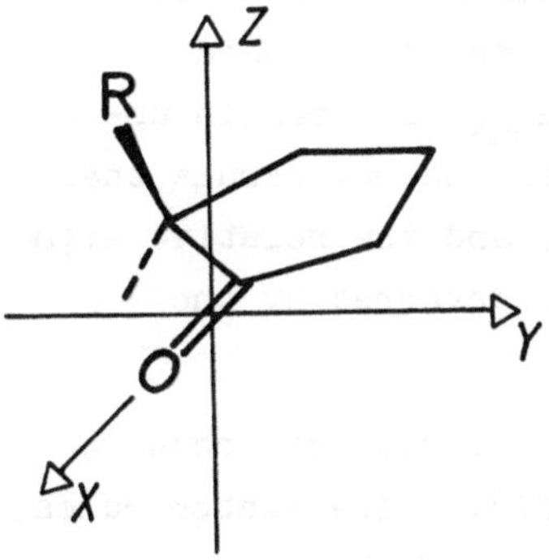

Figure 23a

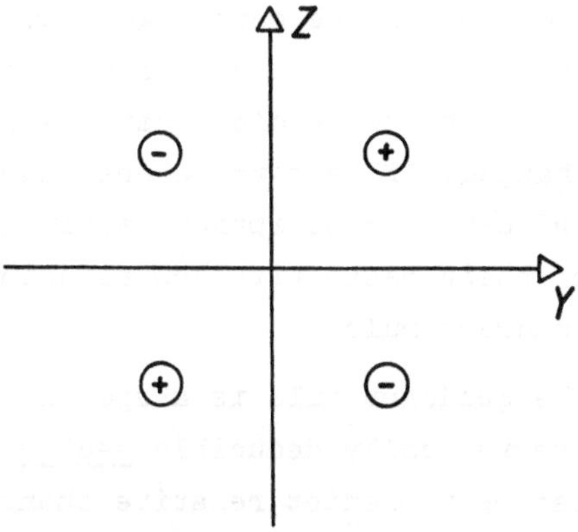

Figure 23b

$$n' = n + \lambda\pi$$

By first-order perturbation theory

$$\lambda = \frac{\langle n|V_R|\pi\rangle}{\Delta E_{n\pi}}$$

where V_R is the potential of the substituent R. The rotatory strength for the $n \rightarrow \pi^*$ transition becomes

$$\mathbb{R}_{n\pi^*} = \lambda\langle\pi|\vec{r}|\pi^*\rangle\langle\pi^*|\vec{m}|n\rangle$$

For a given phase of the MO's the sign of $\mathbb{R}_{n\pi^*}$ will depend on λ. Suppose V_R is everywhere positive in space, corresponding to the potential of the incompletely shielded nuclei of the substituent (a methyl group, for instance): The sign of the matrix element $\langle n|V_R|\pi\rangle$ will vary with the position of R as the sign of the product $(n\cdot\pi)$, or $(y\cdot z)$, at the position of R. This leads to a quadrant rule, as depicted in Figure 23b. If we have just one substituent in a given position of cyclopentanone (and if we assume the ring to be planar, which in fact it is not), this result is trivial: By moving R around the C-O bond we merely go from one enantiomer to the other. However, if we move R from α to β position in the same quadrant, we predict that the sign of $R_{n\pi^*}$ will remain unchanged. If we have several substituents, we may assume their influence to be approximately additive, and the relative sign of their respective contributions to be governed by the quadrant rule.

The quadrant rule is a special case of a series of group theoretically deducible sector rules [20]. These sector rules can only predict relative changes in sign of the rotatory strength. To predict absolute signs, either empirical calibrations, or more elaborate computations are necessary (see Case 3).

Case 2:

We consider a molecule as composed of two identical monomers 1 and 2. We suppose that these monomers themselves are optically inactive, but that they are coupled in such a way that the dimer is optically active. We suppose furthermore that the splitting of the two degenerate electric dipole-allowed longest-wavelength transitions may be interpreted by the dipole-dipole approximation [21] (Figure 24).

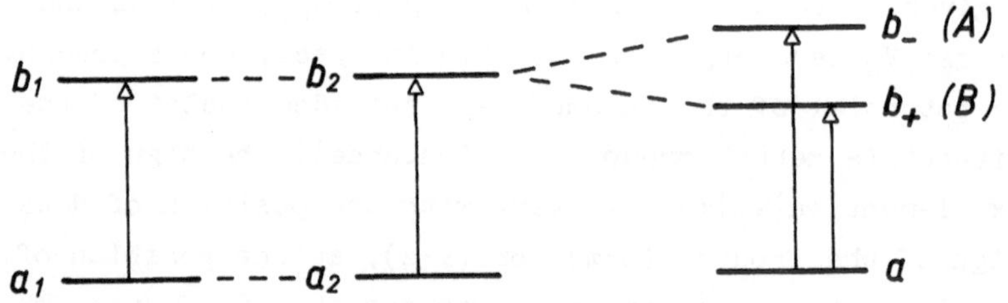

Figure 24

Then it may be shown that to a degree of approximation which we presently shall discuss, the rotatory strength of the two longest-wavelength transitions in the dimer is given by

$$\mathbb{R}_{ab} = C\, \vec{R}_{12} \cdot (\vec{\mu}_2 \times \vec{\mu}_1)$$

where $\vec{R}_{12} = \vec{R}_2 - \vec{R}_1$ is the difference of the position vectors of the monomers with respect to some molecule-fixed origin, and $\vec{\mu}_1$ and $\vec{\mu}_2$ are the electric dipole transition moments of the monomers: $\vec{\mu}_1 \equiv \langle a_1|\vec{R}|b_1\rangle$, $\vec{\mu}_2 \equiv \langle a_2|\vec{R}|b_2\rangle$. C is a positive constant. As an example we choose 2-2'-diamino-6-6'-dimethyl-biphenyl (Figure 25). In a simplifying manner we consider the molecule for our purposes to be represented by two coupled aniline chromophores. We neglect the influence of the methyl substituants and of the bond connecting the rings. We know from

Righthanded Chirality

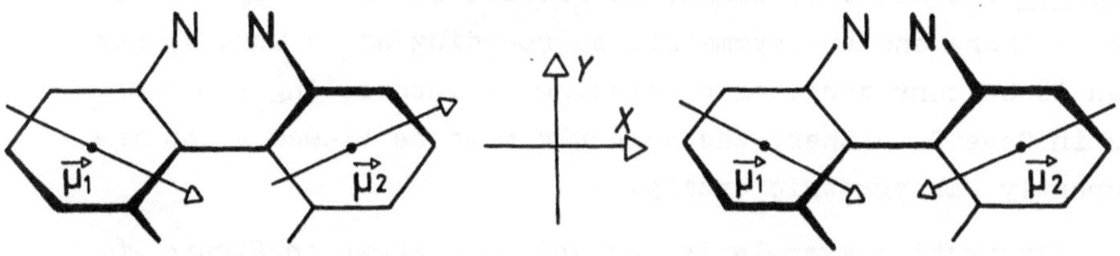

First transition $a \rightarrow b_+$
Lower energy
x,z-polarized
Symmetry B under C_2

Second transition $a \rightarrow b_-$
Higher energy
y-polarized
Symmetry A under C_2

Figure 25

experiment that in aniline the longest-wavelength transition is polarized perpendicularly to the C-N bond [22]. From Figure 25 we find by the coupled oscillator model applied to the composite chromophore of C_2-symmetry and righthanded chirality, in the order of decreasing wavelength (see also Figure 24):

First transition B-polarized; $\mathbb{R}_{ab+}$ positive.
Second transition A-polarized; $\mathbb{R}_{ab-}$ negative.

This appears to agree with experiment [22].

In our simplified approach we have localized the transition moments at the geometric centers of the benzene rings. This is a point of arbitraryness which must be dealt with. In the present case this choice appears to be admissible, but in general, if the dimensions of the monomers are comparable to the distance between them, then it is not at all obvious where we should localize these transition moments. There is, in fact, only <u>one</u> point for each monomer where the corresponding electric transition moment may be localized as a point-dipole, and for which the formula

$$\mathbb{R}_{ab} = C\, \vec{R}_{12} \cdot (\vec{\mu}_2 \times \vec{\mu}_1)$$

is exact, otherwise additional terms appear which may be important and may well even make opposite contributions in sign.

<u>Case 3:</u> The molecule cannot be subdivided into a symmetric chromophore and an asymmetric surrounding as in Case 1, nor can it be considered as consisting of interacting subgroups as in Case 2. Rather, the molecule must be viewed as an inherently dissymmetric entity.

An illustrative example is the molecule shown in Figure 26, which displays a strong optical activity and long-wavelength Cotton effects of opposite sign at 294 nm and 263 nm [23]. If we apply the procedure of Case 2 indistinctly, localizing the electric transition moments $\vec{\mu}_1$ and $\vec{\mu}_2$ in the center of

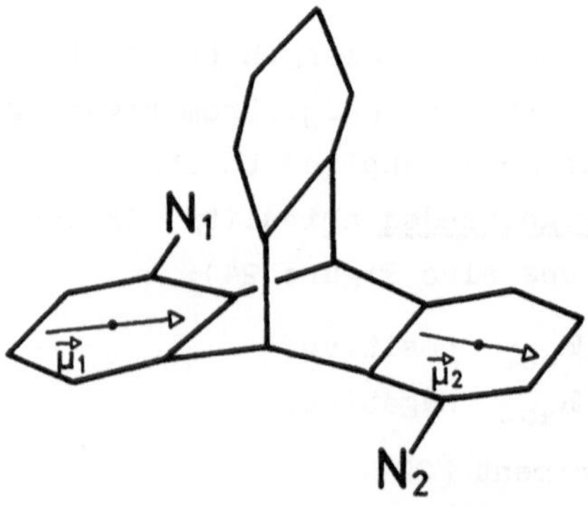

<u>Figure 26</u>

the benzene rings 1 and 2, pointing perpendicularly to the C-N bonds, the computed rotatory strength vanishes, because $\vec{\mu}_1$, $\vec{\mu}_2$ and $\vec{R}_{12}$ are coplanar. This implies that the transition moments should be pointed and localized differently. This must be done cautiously, or it may easily lead to wrong predictions. In other words, a more accurate assessment of the quantity Im $\left\{\langle\psi_a|\vec{R}|\psi_b\rangle\langle\psi_b|\vec{M}|\psi_a\rangle\right\}$ is necessary [24].

A possible procedure, actually leading to correct predictions with regards to order of magnitude and sign of the Cotton effects, is the following [10,25]:

1) Compute the SCF ground state (Section IV.1.) and the lowest excited states by single-excitation CI (Section III.2.), assuming local σ-π separation in all three benzene rings and invoking the PPP approximation (Section III.3.). The nitrogen atoms of the substituents (Figure 26) are considered to be of the pyrrhole-type and to cintribute two (pseudo-) π electrons each. We thus treat the molecule as an inherently dissymetric (pseudo-) 22 π electron system. The γ_{pq} integrals between all eighteen carbon atoms and the two nitrogen atoms enter the computation. Only the resonance integrals β_{pq} between nearest neighbors within a conjugated subunit (N_1-benzene ring 1; N_2-benzene ring 2; benzene ring 3) are taken into account. The core matrix thus has the aspect given in Figure 27.

2) From the semiempirical wavefunctions the rotatory strength of the longest wavelength transitions are then evaluated.

In practice it proves necessary to compute the electric dipole transition moment in the dipole velocity form. For exact eigenfunctions ψ_a and ψ_b of the same Hamiltonian we have the identity (see p. 39 for definition of $\vec{R}$):

$$\langle\psi_a|\vec{R}|\psi_b\rangle = \frac{e\hbar^2}{m(E_b-E_a)}\langle\psi_a|\vec{\nabla}|\psi_b\rangle$$

where $\vec{\nabla} \equiv \sum_{\mu} \vec{\nabla}_\mu$. For the rotatory strength we then obtain

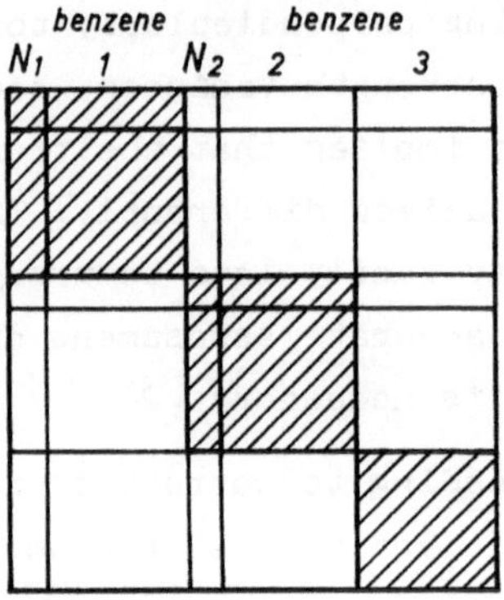

Figure 27

$$\mathbb{R}_{ab} = \frac{e^2\hbar^2}{2m^2c(E_b-E_a)} \operatorname{Im}\left\{\langle\psi_a|\vec{\nabla}|\psi_b\rangle\langle\psi_b|\vec{L}|\psi_a\rangle\right\}$$

where $\vec{L} \equiv \sum_{\mu} \vec{\ell}_{\mu}$.

Expressing the wavefunctions in terms of configurational functions

$$\psi_a \approx \Phi_0 \quad ; \quad \psi_b \approx \sum_{ik} B_i^k \Phi_i^k$$

we find (see Section III.5.):

$$\langle\psi_a|\vec{\nabla}|\psi_b\rangle \approx \sqrt{2}\sum_{ik} B_i^k \langle\varphi_i|\vec{\nabla}|\varphi_k\rangle \qquad \text{and}$$

$$\langle\psi_b|\vec{L}|\psi_a\rangle \approx \sqrt{2}\sum_{ik} B_i^k \langle\varphi_k|\vec{\ell}|\varphi_i\rangle$$

The configurational coefficients B_i^k are assumed real, likewise the SCF-MO's φ_i. As $\vec{\ell} \equiv -i\hbar\, \vec{r} \times \vec{\nabla}$ and as $\langle\varphi_k|\vec{r}\times\vec{\nabla}|\varphi_i\rangle = - \langle\varphi_i|\vec{r}\times\vec{\nabla}|\varphi_k\rangle$, we finally obtain

$$\mathbb{R}_{ab} = \frac{e^2\hbar^3}{m^2c(E_b-E_a)} \sum_{ik}\sum_{i'k'} B_i^k B_{i'}^{k'} \langle\varphi_i|\vec{\nabla}|\varphi_k\rangle\langle\varphi_{i'}|\vec{r}\times\vec{\nabla}|\varphi_{k'}\rangle$$

The approximations inherent in this expression originate in the approximate nature of ψ_a and ψ_b.

Returning to our example, the φ_i and B_i^k are computed as described under 1). The matrix elements $\langle\varphi_i|\vec{\nabla}|\varphi_k\rangle$ and $\langle\varphi_i|\vec{r}\times\vec{\nabla}|\varphi_k\rangle$ are numerically evaluated without further approximations. They reduce to integrals between atomic orbitals which, in the case of Slater orbitals, boil down to the evaluation of linear combinations of overlap integrals [25]. The matrix elements of $\vec{r}\times\vec{\nabla}$ depend on the origin of the coordinate system; the matrix elements of $\vec{\nabla}$ do not. It may be shown that R_{ab}, as computed above, likewise is origin-independent, which of course it should be. However, as mentioned, the electric dipole transition moment must be computed in the dipole velocity form [26].

Using standard PPP parametrization [22] and taking into account 99 singly excited configurations we obtain for the longest-wavelength Cotton effects of our triptycene derivative (Figure 26):

$$\psi_0 \rightarrow \psi_1 : \lambda = 291 \text{ nm}, \quad R_{01} = -0.44\cdot 10^{-38} \text{ cgs}$$

$$\psi_0 \rightarrow \psi_2 : \lambda = 285 \text{ nm}, \quad R_{02} = +0.33\cdot 10^{-38} \text{ cgs}$$

$$\psi_0 \rightarrow \psi_3 : \lambda = 264 \text{ nm}, \quad R_{03} = +0.21\cdot 10^{-38} \text{ cgs}$$

This appears to agree with experiment as to order of magnitude and sign.

2. Selection rules for electrocyclic reactions and cycloaddition reactions

Molecular orbital theory in its simplest form, in particular the EH approximation [15], has provided a brilliant means of rationalizing and interpreting the regularities encountered in concerted organic reactions. The corresponding rules, now generally called Woodward-Hoffmann-rules [27], grew out of the necessity to rationalize empirical evidence such as the following:

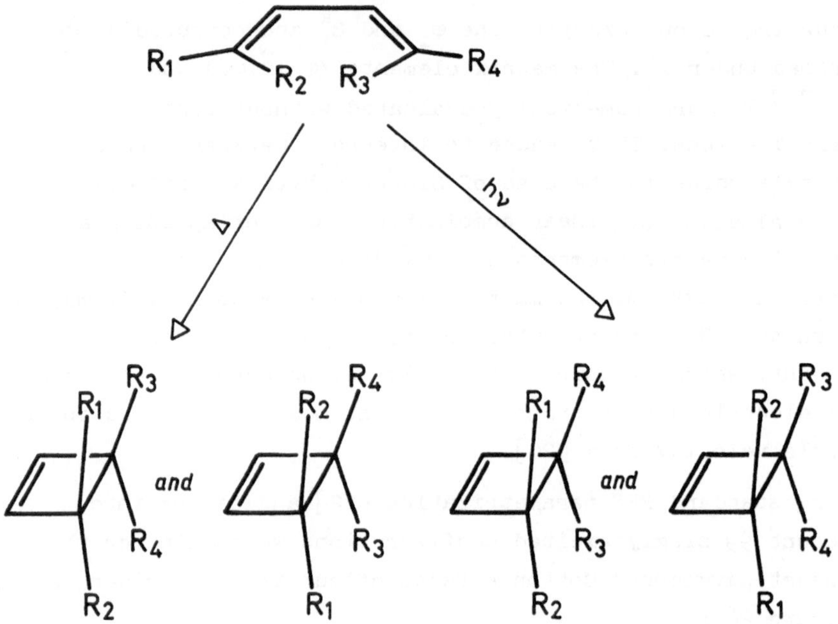

Thermal (Δ) ring closure of butadiene proceeds by a conrotatory movement of the substituents, photochemical (hν) ring closure goes in a disrotatory way. For hexatriene it is the opposite.

Already prior to Woodward's and Hoffmann's systematic investigations, Fukui and Oosterhoff had independently [28] suggested that the course of such reactions might be connected to the symmetry of the highest occupied orbital of the polyene. The $2p_\pi$ AO's of the terminal carbon atoms may be thought of as combining with the proper phase to form a bonding σ orbital. In butadiene in the ground state (thermal path) the highest filled orbital is π_2 (Figure 28), while in the first excited state (photochemical path) it is π_3.

From a more general point of view it became clear that such selection rules could be better interpreted by looking at the

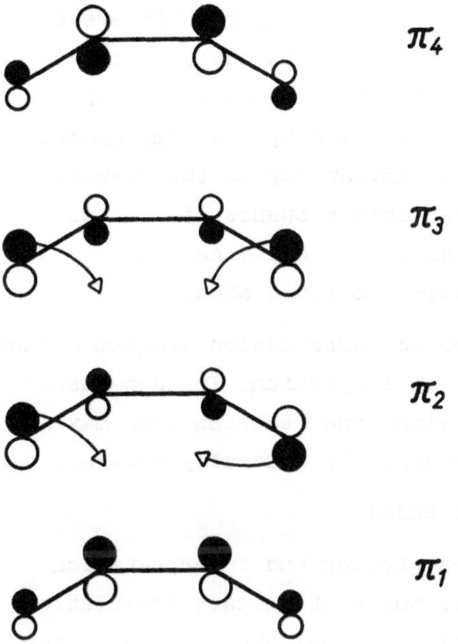

Figure 28

totality of participating electrons [29]. Assuming that the states of the reactant and of the product may be characterized by a common and <u>relevant</u> symmetry element, correlation diagrams can be drawn as shown in Figures 29a,b and 30a,b for electrocyclic reactions, and 31a and 31b for cycloaddition reactions. The following points are of importance:

1) During the transformation at least one element of (overall or local) symmetry is maintained. The states of reactant, product and most plausible transition state may be characterized by it.
2) The symmetry element of importance must bisect the bond(s) which is (are) being formed or broken.
3) In correlating the states of reactant and product the "noncrossing rule" holds: Lines correlating states of same symmetry may not cross.

4) For simplicity it is assumed that the states of interest can be described by single configurations, defined in a set of appropriate one-electron MO's. The one-electron states are of course also caracterized by the same symmetry element(s) and correlate in a similar way as the overall many-electron states. The energetic sequence of the one-electron states and their occupation determines the energetic sequence of the many-electron states.

5) There are cases where the correct correlation diagram cannot be established unambiguously by inspection. In such cases a series of EH calculations along the reaction path may illustrate how the molecular orbitals gradually evolve.

Two remarks must immediately be added:

- The present MO theoretical description and interpretation does not automatically lead to the most compact formulation of the selection rules, as the practical chemist seeks them. On this question there also exists a vast amount of literature [30].

- The simplicity of the elementary molecular orbital approach invites further refinement. Until now it appears that more sophisticated treatments basically confirm the results obtained from the simpler picture [31].

We now turn to elementary examples:

Figure 29a illustrates the electrocyclic reaction of butadiene over a conrotatory path. The symmetry element C_2 is maintained throughout and the one-electron states correlate as shown. The transformation properties of the corresponding many-electron states, denoted by big letters A (symmetric) or B (antisymmetric), is immediately deducible from the symmetry, designated by small letters a or b, of the MO's occupied by electrons participating in the reaction. In the thermal case we see that the ground state (ground configuration) of butadiene correlates with the ground state (ground configuration)

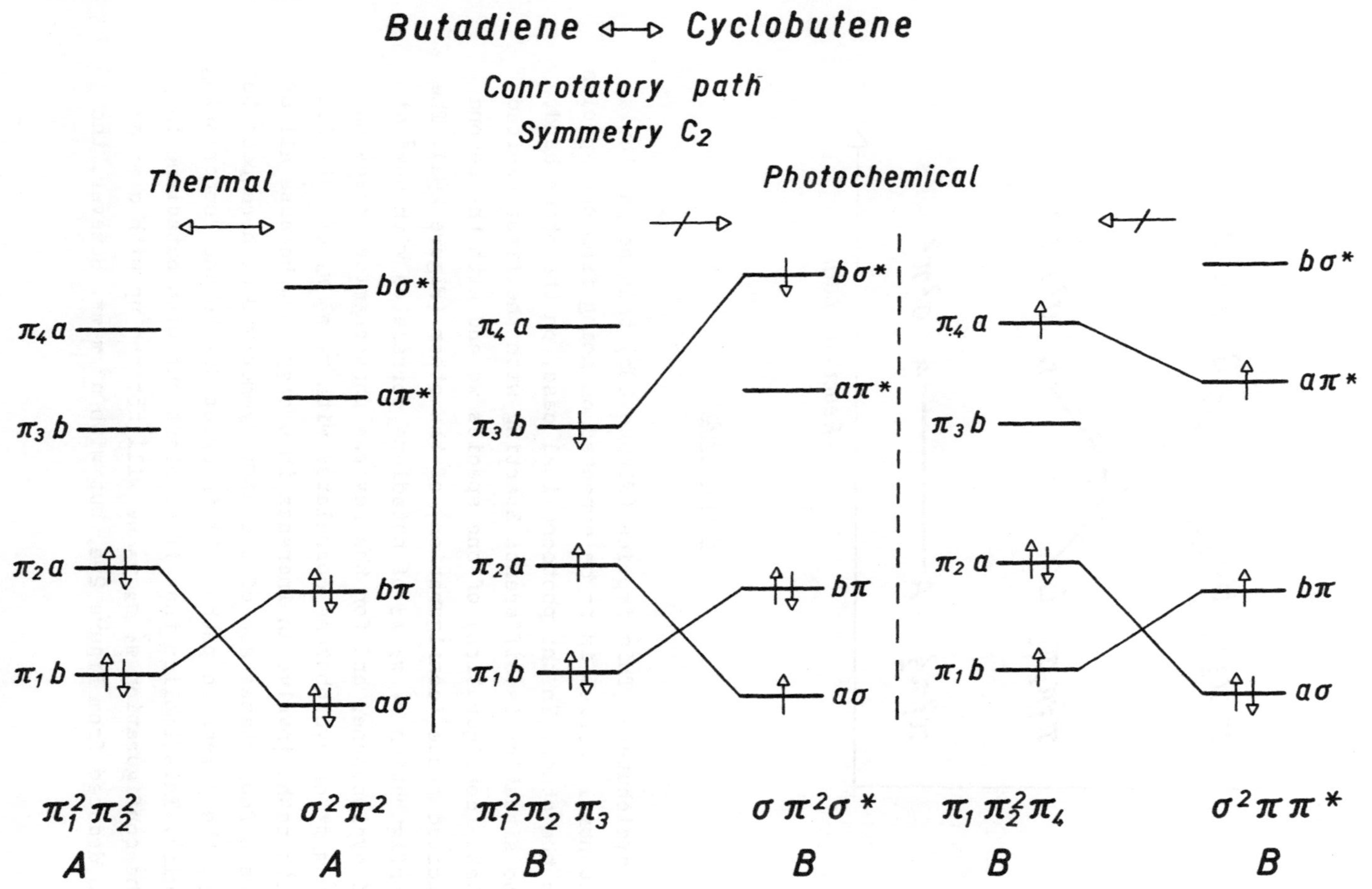

Butadiene ↔ Cyclobutene
Conrotatory path
Symmetry C_2
Thermal
Photochemical
$\pi_4 a$
$\pi_3 b$
$\pi_2 a$
$\pi_1 b$
$b\sigma^*$
$a\pi^*$
$b\pi$
$a\sigma$
$\pi_1^2 \pi_2^2$
A
$\sigma^2 \pi^2$
A
$\pi_1^2 \pi_2 \pi_3$
B
$\sigma \pi^2 \sigma^*$
B
$\pi_1 \pi_2^2 \pi_4$
B
$\sigma^2 \pi \pi^*$
B

Figure 29a

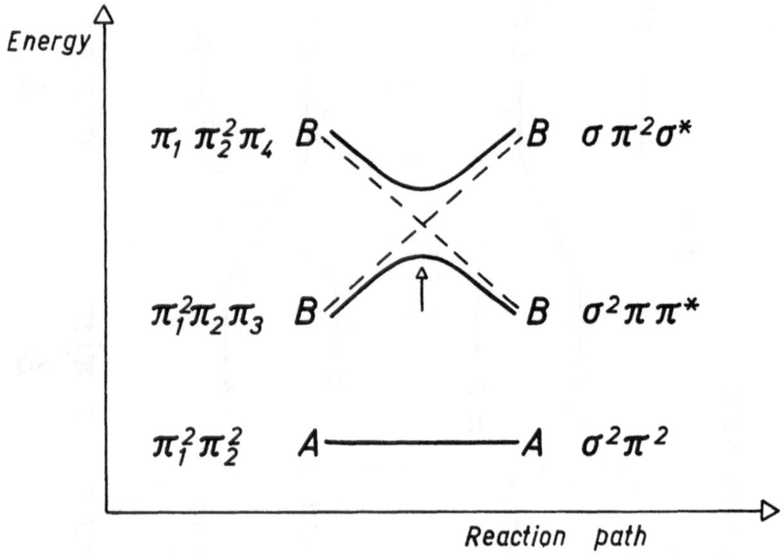

Figure 29b

of cyclobutene. This implies (Figure 29b) that no particular potential barrier is to be expected on going from one species to the other. In the photochemical case, on the other hand, the situation is different: Starting with the first excited state (configuration) of one species we end with the second excited state (configuration) of the other (Figure 29a). The configuration $\pi_1^2\ \pi_2\ \pi_3$ of butadiene correlates with $\sigma\ \pi^2\ \sigma^*$ of cyclobutene, and for the reverse photochemical reaction $\sigma^2\ \pi\ \pi^*$ of cyclobutene correlates with $\pi_1\ \pi_2^2\ \pi_4$ of butadiene. Both paths involve an increase in energy. Now, because all of these four states are of the same symmetry B with respect to C_2, the adiabatic paths will interact due to the "noncrossing rule". This implies that if we start out with butadiene in the configuration $\pi_1^2\ \pi_2\ \pi_3$ we will not end up with $\sigma\ \pi^2\ \sigma^*$ as deduced from Figure 29a, but with $\sigma^2\ \pi\ \pi^*$. However, the

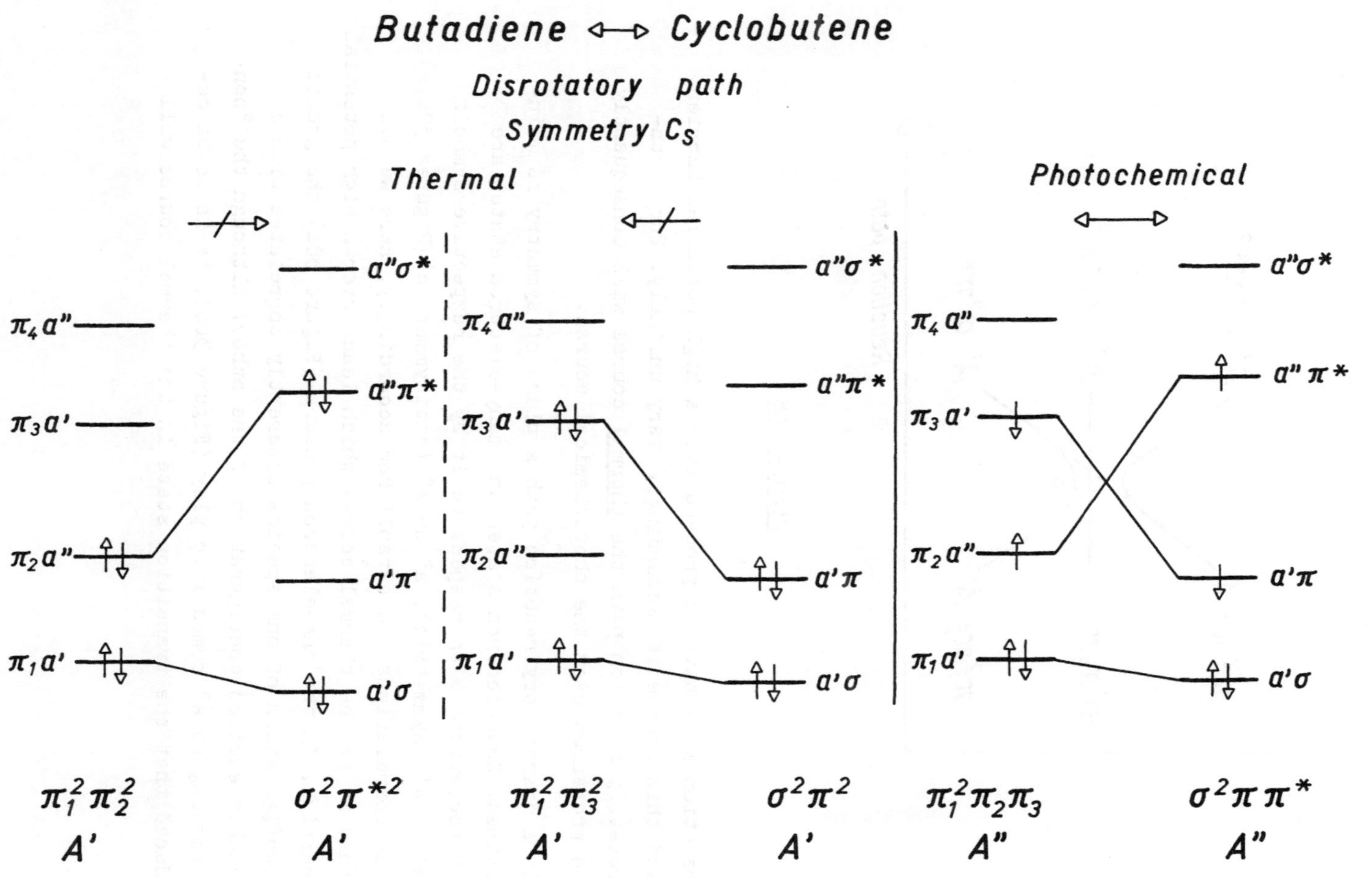
Butadiene ⟷ Cyclobutene
Disrotatory path
Symmetry C_s
Thermal
Photochemical
$\pi_4 a''$
$\pi_3 a'$
$\pi_2 a''$
$\pi_1 a'$
$a''\sigma^*$
$a''\pi^*$
$a'\pi$
$a'\sigma$
$\pi_1^2\pi_2^2$
A'
$\sigma^2\pi^{*2}$
A'
$\pi_1^2\pi_3^2$
A'
$\sigma^2\pi^2$
A'
$\pi_1^2\pi_2\pi_3$
A''
$\sigma^2\pi\pi^*$
A''

Figure 30a

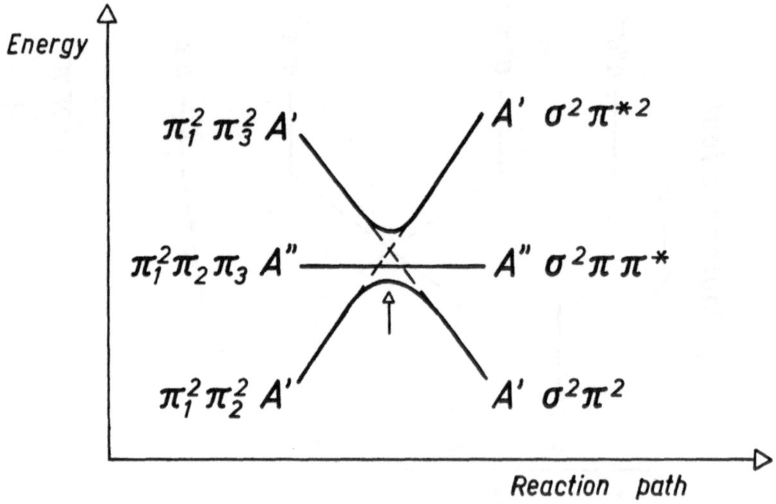

Figure 30b

reaction will have to proceed over a high potential barrier, and this course is accordingly very unlikely. For a conrotatory reaction path the thermal course will consequently be preferred over the photochemical course.

In a disrotatory reaction path a plane of symmetry is maintained. One-electron states and many-electron states are characterized with respect to it by the respective symbols a' or A' (symmetric), a" or A" (antisymmetric) (Figures 30a,b). The correlations are established accordingly. Here we see that it is the thermal course which leads over a high potential barrier. In the one-electron picture (Figure 30a) the ground configuration of one species apparently correlates with a doubly excited configuration of the other. Although the "non-crossing rule" comes into play (Figure 30b), it is to be deduced that the transition state in the thermal course will nonetheless be of relatively very high energy. On the other hand, in the photochemical case, the lowest singly excited configuration of reactant always correlates with the same singly excited configuration of the product. In the disrotatory reaction path the photochemical course will consequently be favored over the thermal course.

We now turn to cycloaddition reactions and consider the example of the addition of ethylene and butadiene to give cyclohexene. The course of this reaction proceeds in such a way that a plane of symmetry is maintained.

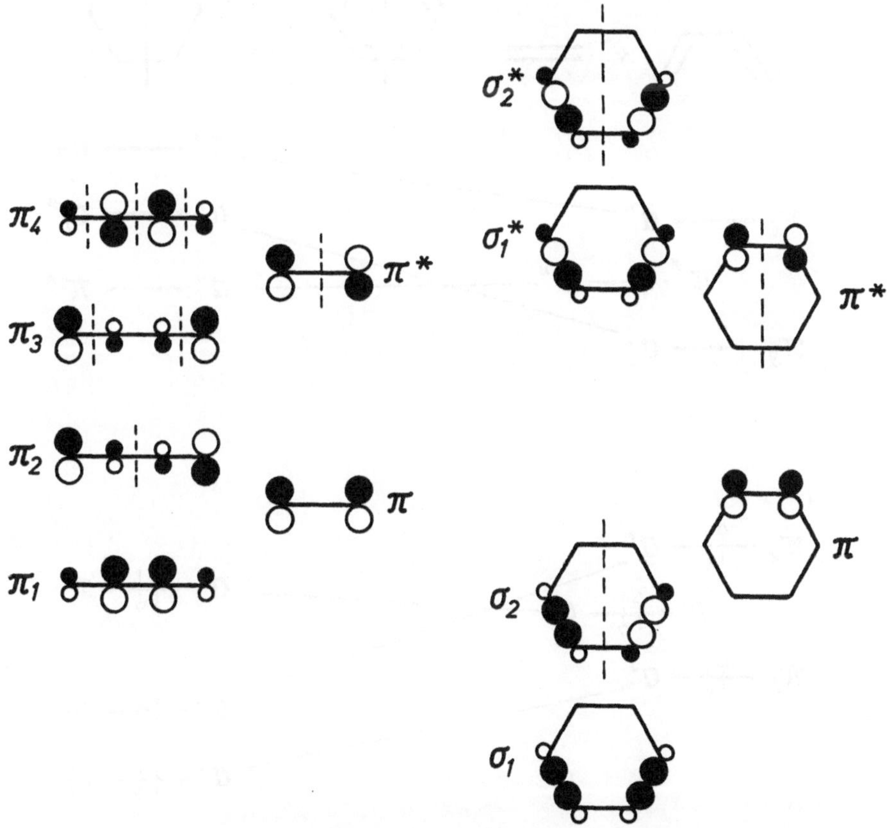

Figure 31a

In accord with group theoretic usage we characterize the one-electron states by a' and a" with respect to the relevant plane of symmetry, as in the previous example. Figure 31a shows at left the three highest occupied orbitals of the reactants butadiene+ethylene in the proper relative energetic sequence π_1(a'), π(a'), π_2(a") and the three lowest unoccupied orbitals in the corresponding relative sequence π_3(a'), π^*(a"), π_4(a"). In Figure 31a at left the three highest occupied and three lowest unoccupied orbitals of the product cyclohexene are drawn

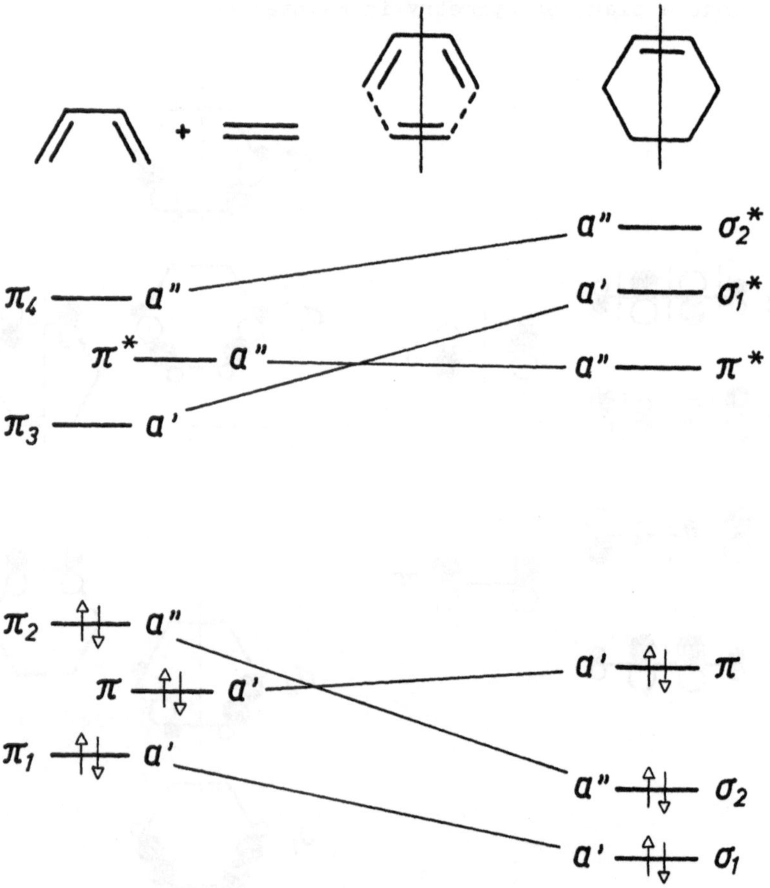

Figure 31b

in a simplified way, also in the proper energetic sequence $\sigma_1(a')$, $\sigma_2(a'')$, $\pi(a')$; $\pi^*(a'')$, $\sigma_1^*(a')$, $\sigma_2^*(a'')$. The corresponding correlation diagram for the one-electron states is shown in Figure 31b. Considering the occupation of the orbitals, we see that the ground configuration of the reactant correlates with the ground configuration of the product, making the reaction <u>thermally allowed</u>.

The photochemical course, starting from the lowest excited configuration $\pi_1^2\ \pi^2\ \pi_2\ \pi_3$ of the reactants, correlates with the higher excited configuration $\sigma_1^2\ \sigma_2^2\ \pi\ \sigma_1^*$ of the product and is therefore energetically unfavorable. The question is of interest, if the reaction could not proceed from the photoexcited ethylene, instead of the photoexcited butadiene. The configuration $\pi_1^2\ \pi\ \pi_2^2\ \pi^*$ indeed correlates with the lowest excited configuration of the product, $\sigma_1^2\ \sigma_2^2\ \pi\ \pi^*$. However, it is possible that as soon as the reactant molecules interact, the state $\pi_1^2\ \pi\ \pi_2^2\ \pi^*$ internally converts by vibronic coupling to the lower excited state $\pi_1^2\ \pi^2\ \pi_2\ \pi_3$, thereby impeding the further course of the reaction.

For a discussion of other concerted organic reactions, such as sigmatropic or cheletropic reactions, the reader is referred to the literature [27,30].

<u>Exercise:</u> Draw the state correlation diagram for the cycloaddition reaction ethylene+butadiene, based on Figure 31b.

3. <u>Molecular orbital theory with periodic (cyclic) boundary conditions</u>

This form of molecular orbital theory provides a means of studying in the tight-binding approximation the electronic structure of polymers built from sequentially repeating subunits [32]. It is assumed that, neglecting end effects, cyclic boundary conditions may be applied. We consider a polymer consisting of N monomers and assume that per monomer ν valence orbitals (atomic orbitals) have to be taken into account. The number of electrons per monomer is ν_e. Furthermore, we denote the monomers by the indices p, p', q, q', the AO's within the monomers by s, s', t, t'. We characterize the polymer MO's by double indices jm, j'm', ℓn, ℓ'n', where j, j', ℓ, ℓ' designate the symmetry of the MO under the cyclic point group C_N.

The s th atomic orbital within the p th monomer is written χ_{ps}. Let us choose the origin of the molecule-fixed coordinate system to coincide with the center of a particular AO, $\chi_s \equiv \chi_s(\vec{r})$. The corresponding AO $\chi_s(\vec{r}-\vec{a})$ will be centered in the point $\vec{a}$, $\chi_s(\vec{r}-2\vec{a})$ will be centered in the point $2\vec{a}$, etc. If $\vec{a}$ is the primitive translation vector of a linear polymer we may write

$$\chi_s(\vec{r}) = \chi_{0s}$$

$$\chi_s(\vec{r}-\vec{a}) = \chi_{1s}$$

$$\chi_s(\vec{r}-2\vec{a}) = \chi_{2s}$$

.....

$$\chi_s(\vec{r}-p\vec{a}) = \chi_{ps}$$

Considering the periodic boundary conditions or formal cyclic symmetry we in general write a molecular orbital, extending over the whole polymer, as [33]:

$$\varphi_{jm}(\vec{r}) = \sum_{p=1}^{N} \sum_{s=1}^{\nu} \omega^{jp} c_{jm,s} \chi_{ps}$$

or in bracket notation

$$|jm\rangle = \sum_{p} \sum_{s} \omega^{jp} c_{jm,s} \, |ps\rangle$$

where $\omega \equiv \exp(2\pi i/N)$. In a given polymer the χ_{ps} will have to be appropriately chosen, so as to conform to the particular (for instance helical) geometry. From the above relations one finds

$$\varphi_{jm}(\vec{r}+\vec{a}) = \sum_{p=1}^{N} \sum_{s=1}^{\nu} \omega^{jp} c_{jm,s} \chi_{(p-1)s}$$

$$= \omega^{j} \sum_{p=1}^{N} \sum_{s=1}^{\nu} \omega^{j(p-1)} c_{jm,s} \chi_{(p-1)s}$$

which, due to the cyclic properties

$$= \omega^{j} \varphi_{jm}(\vec{r})$$

and consequently

$$\varphi_{jm}(\vec{r}+N\vec{a}) = \omega^{jN} \varphi_{jm}(\vec{r}) = \varphi_{jm}(\vec{r})$$

The function must indeed have the same value after N elementary translations.

<u>The Hückel/Extended Hückel approximation:</u>

See Chapter II and Section V.1.

We define symmetry orbitals (in bracket notation):

$$|js\rangle = \sum_{p=1}^{N} \omega^{jp} \, |ps\rangle$$

$$|j's'\rangle = \sum_{p'=1}^{N} \omega^{j'p'} \, |p's'\rangle$$

$j, j' = 1 \ldots N$
or equivalently, for N even:
$j, j' = 0, \pm 1, \ldots \pm (N/2-1), N/2$
for N odd:
$j, j' = 0, \pm 1, \ldots \pm (N-1)/2$

and express the eigenvalue equation in the basis of these orbitals

$$\left| \langle js|h|j's'\rangle - \epsilon \langle js|j's'\rangle \right| = 0$$

This is equal to

$$\left| \sum_{p=1}^{N} \sum_{p'=1}^{N} \omega^{-jp+j'p'} \left\{ \langle ps|h|p's'\rangle - \epsilon\langle ps|p's'\rangle \right\} \right| = 0$$

The indices s, s' go from 1 to ν. It may be shown [34] that the elements of this secular determinant vanish, unless $j = j'$. The eigenvalue problem, originally of order $N\cdot\nu$, thus factorizes into separate equations of order ν for every value of j.

Example: A cyclic chain of N "ethylene" molecules (see Figure 32). We have s, s' = 1,2

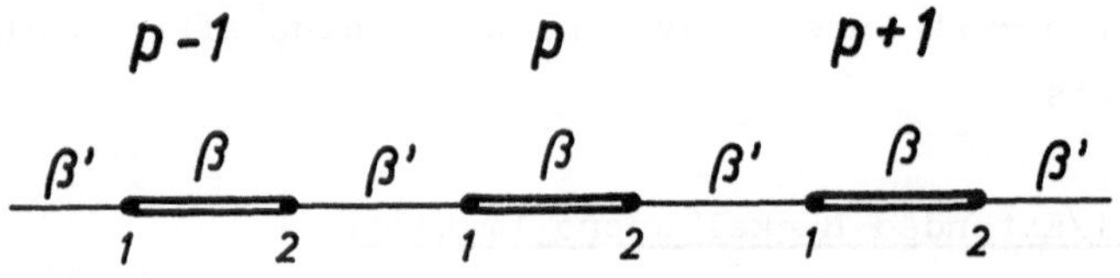

Figure 32

Invoking the ZDO and nearest-neighbor approximation, as in ordinary Hückel theory, we find

$$\langle p1|h|p1\rangle = \langle p2|h|p2\rangle = \alpha$$
$$\langle p1|h|p2\rangle = \beta$$
$$\langle p1|h|(p-1)2\rangle = \langle p2|h|(p+1)1\rangle = \beta'$$

Adding the contributions for p = 1,2 ... N and then dividing every element by N we obtain for the secular equation

$$\begin{vmatrix} \alpha - \varepsilon & \beta + \omega^{-j}\beta' \\ \beta + \omega^{j}\beta' & \alpha - \varepsilon \end{vmatrix} = 0$$

Leading to the solutions

$$\varepsilon = \alpha \pm \left\{\beta^2 + \beta'^2 + 2\beta\beta' \cos \frac{2\pi j}{N}\right\}^{1/2}$$

For $\beta' = 0$, the formula reduces to the Hückel energy levels of independent ethylene molecules. For $\beta = \beta'$ we get

$$\varepsilon = \alpha \pm \beta \left\{2\left(1+\cos \frac{2\pi j}{N}\right)\right\}^{1/2} = \alpha \pm 2\beta \cos \frac{\pi j}{N}$$

With N=3 we correctly obtain the Hückel energy levels of benzene (see page 10), j taking on the values 0, ± 1.

The SCF formulation:

See Sections IV.1. and IV. 3 (Attention: The symbol N there and here has a different meaning.)

The Hartree-Fock equation in the basis of symmetry orbitals may be expressed as

$$\left| \langle js|F|j's'\rangle - \varepsilon\langle js|j's'\rangle \right| = 0$$

$$\langle js|F|j's'\rangle$$

$$= \langle js|h|j's'\rangle + \sum_{\ell=1}^{N} \sum_{n=1}^{\nu_e/2} \left\{2 \langle js\ \ell n|j's'\ \ell n\rangle - \langle js\ \ell n|\ell n\ j's'\rangle\right\}$$

The summation goes over all occupied polymer orbitals. Here we assume that there is no overlapping of the energies of bands for which $n \leq \nu_e/2$ and of bands for which $n > \nu_e/2$. We admit that we have the situation shown in Figure 33a and not the one illustrated by Figure 33b. A polymer MO $|\ell n\rangle$ is written

$$|\ell n\rangle = \sum_q \sum_t \omega^{\ell q} c_{\ell n,t}\ |qt\rangle$$

The symmetry index ℓ - like j - takes on the values $\ell = 1, 2 \ldots N$ or, equivalently, $\ell = 0, \pm 1, \ldots \pm(N/2-1), N/2$. It may be shown [34] that matrix elements of the Fock operator between symmetry orbitals vanish, unless $j' = j$. With $s, s' = 1, 2 \ldots \nu$, the Fock matrix consequently factorizes into N submatrices of order ν. Expanding the Fock operator we find:

$$\langle js|F|js'\rangle = \Bigg[\sum_{p=1}^{N} \sum_{p'=1}^{N} \omega^{j(p'-p)} \langle ps|h|p's'\rangle + \sum_{\ell=1}^{N} \sum_{n=1}^{\nu_e/2} \sum_{t=1}^{\nu} \sum_{t'=1}^{\nu} c^*_{\ell n,t} c_{\ell n,t'} \sum_{p=1}^{N} \sum_{q=1}^{N} \sum_{p'=1}^{N} \sum_{q'=1}^{N} \omega^{j(p'-p)+\ell(q'-q)} \cdot \left\{ 2\langle ps\; qt|p's'\; q't'\rangle - \langle ps\; qt|q't'\; p's'\rangle \right\} \Bigg]$$

This formula is exact within the Hartree-Fock approximation. It is indeed rather cumbersome to evaluate this expression for large N, in particular the two-electron part. The fall-off

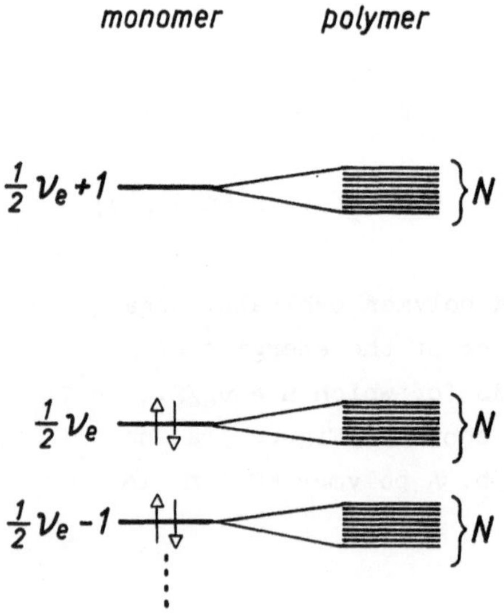

Figure 33a

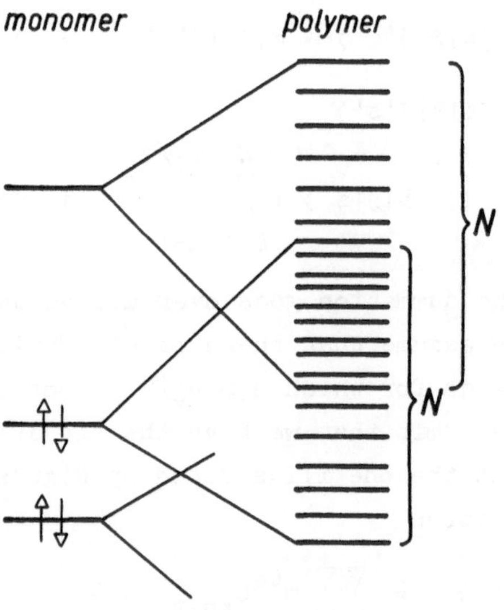

Figure 33b

of the corresponding integrals with distance may be taken into account to simplify it, as indicated in [34].

Within the frame of a semiempirical approach the ZDO approximation is useful (see Section IV.2). The electron repulsion integrals $\langle ps\ qt|p's'\ q't'\rangle$ are then neglected, except when $ps = p's'$ and $qt = q't'$. Similarly only integrals of the form $\langle ps\ qt|q't'\ p's'\rangle$ are retained for which $ps = q't'$ and $qt = p's'$. We remember that s and s' are fixed indices for a given matrix element, while t and t' are running indices. The two-electron part then reduces

for the <u>diagonal</u> element $\langle js|F|js\rangle$ to:

$$\sum_{\ell=1}^{N} \sum_{n=1}^{\nu_e/2} \Big\{ 2 \sum_{t} c^*_{\ell n,t} c_{\ell n,t} \sum_{p}\sum_{q} \langle ps\ qt|ps\ qt\rangle - c^*_{\ell n,s} c_{\ell n,s} \sum_{p}\sum_{q} \omega^{(j-\ell)(q-p)} \langle ps\ qs|ps\ qs\rangle \Big\}$$

and for the <u>nondiagonal</u> element $\langle js|F|js'\rangle$ to:

$$-\sum_{\ell=1}^{N} \sum_{n=1}^{\nu_e/2} c^*_{\ell n,s}\ c_{\ell n,s'} \sum_{p}\sum_{q} \omega^{(j-\ell)(q-p)} \langle ps\ qs'|ps\ qs'\rangle$$

<u>The computation of optical properties</u> [34]: The CI matrix also factorizes according to the irreducible representations of the group C_N (for N even):

$$A,\ E_{+1},\ E_{-1},\ \ldots\ldots\ E_{+j},\ E_{-j}\ \ldots\ldots\ E_{+(N/2-1)},\ E_{-(N/2-1)},\ B$$

One starts from a total of $N\cdot\nu$ polymer MO's, of which $N\cdot\nu_e/2$ are filled. Consequently $N\cdot\mathcal{N}$ possible singly excited configurations may be constructed, where

$$\mathcal{N} = N \cdot \frac{\nu_e}{2}\left(\nu - \frac{\nu_e}{2}\right)$$

With respect to their symmetry, these singly excited configurations are evenly distributed over all the irreducible representations, as long as we assume a situation as shown in Figure 33a. Consequently there should be $\mathcal{N}$ singly excited configurations belonging to each of the N representations A(j=0), B (j=N/2) and sub-representations E_{+j}, E_{-j}. As the dimension of the polymer grows, the computational labor of diagonalizing these N large matrices will rapidly become immense. To limit the expense in calculating optical spectra, one should then give priority to the excited states to which transitions from the ground state are electric dipole allowed. The corresponding selection rules are therefore of immediate interest. We assume that higher retardation effects may be neglected. To derive the selection rules it is essential to distinguish between the formal cyclic symmetry C_N - which we assume always to apply in the sense described - and the actual geometry.

One finds:

a) For a linear geometry transitions are allowed when j' = j. This implies that there is only one CI submatrix of dimension $\mathcal{N}$ between singly excited configurations belonging to the irreducible representation A to be considered.

b) For a cyclic geometry transitions parallel to the symmetry axis of rotation are allowed when j' = j. In-plane transitions are allowed when j' = j±1. This implies considering the CI matrices belonging to the irreducible representations A, E_{+1} and E_{-1}.

c) For a helical geometry the selection rule j' = j holds in the case of transitions parallel to the helical axis. For components polarized perpendicularly one obtains j' = j±M, where M = N/Q and Q is the number of monomers per helical turn. In general, Q will not be an integer. However, M must be an integer. Therefore a clear-cut selection rule will only hold for values of N which correspond to one or several translational identity periods along the axis of the helix. The CI matrices to be considered will belong to the irreducible representations A and E_{+M}, E_{-M}.

Some References

[1] M. Born and R. Oppenheimer, Ann. d. Physik 84, 457 (1927).

[2] H. Eyring, J. Walter and G.E. Kimball, "Quantum Chemistry", John Wiley, New York 1963.

[3] a) E. Hückel, Z. Physik 70, 204 (1931).

b) R. Daudel, R. Lefebvre and C. Moser, "Quantum Chemistry, Methods and Applications", Interscience, New York 1959.

c) A. Streitwieser, "Molecular Orbital Theory for Organic Chemists", John Wiley, New York 1962.

d) L. Salem, "The Molecular Orbital Theory of Conjugated Systems", Benjamin, Inc., 1966.

e) E. Heilbronner and P.A. Straub, Table of Hückel Molecular Orbitals, Springer-Verlag, 1966.

f) E. Heilbronner and H. Bock, "Das HMO-Modell und seine Anwendung, Grundlagen und Handhabung", Verlag Chemie, Weinheim 1968.

[4] A. Carrington and A.D. McLachlan, "Introduction to Magnetic Resonance", Harper and Row, Hew York 1969, p. 89.

[5] R.G. Parr, "Quantum Theory of Molecular Electronic Structure", Benjamin, Inc., New York 1963.

[6] a) E.U. Condon and G.H. Shortley, "The Theory of Atomic Spectra", Cambridge University Press, Cambridge 1963, p. 169-174.

b) J.C. Slater, "Quantum Theory of Atomic Structure", Vol. I, Mc Graw-Hill, 1960, p. 291-295.

[7] J.A. Pople, Proc. Phys. Soc. A 68, 81 (1955).

[8] R. Pariser and R.G. Parr, J. Chem. Phys. 21, 466, 767 (1953).

[9] W. Moffitt, J. Chem. Phys. 22, 320 (1954).

[10] J. Fiutak, Canad. J. Phys. 41, 12 (1963); see also R.E. Geiger and G. Wagnière, in "Wave Mechanics, the first fifty years", ed. W.C. Price, S.S. Chissick, T. Ravensdale, Butterworths, London 1973, Chap. 18.

[11] a) R.S. Mulliken, J. chim. phys. 46, 497, 675 (1949).

b) C.C.J. Roothaan, Rev. Mod. Phys. 23, 69 (1951).

[12] C. Edmiston and K. Ruedenberg, J. Chem. Phys. 43, S97 (1965).

[13] C.C.J. Roothaan, Rev. Mod. Phys. 32, 179 (1960).

[14] J.A. Pople and R.K. Nesbet, J. Chem. Phys. 22, 571 (1954).

[15] R. Hoffmann, J. Chem. Phys. 39, 1397 (1963); see also W. Hug and G. Wagnière, Tetrahedron 25, 631 (1969).

[16] M. Wolfsberg and L. Helmholz, J. Chem. Phys. 20, 837 (1952).

[17] R.S. Mulliken, J. Chem. Phys. 23, 1833, 1841 (1955).

[18] J.A. Pople, D.P. Santry and G.A. Segal, J. Chem. Phys. 43, S129 (1965); J.A. Pople and G.A. Segal, ibid. 43, S136 (1965); see also M. Jungen, H. Labhart and G. Wagnière, Theoret. Chim. Acta 4, 305 (1966); J.M. Sichel and M.A. Whitehead, Theoret. Chim. Acta 7, 32 (1967); R.J. Wratten, Chem. Phys. Letters 1, 667 (1968).

[19] L. Rosenfeld, Z. Phys. 52, 161 (1929); E.U. Condon, Rev. Mod. Phys. 9, 432 (1937); E.U. Condon, W. Altar and H. Eyring, J. Chem. Phys. 5, 753 (1937).

[20] W. Moffitt, R.B. Woodward, W. Klyne and C. Djerassi, J. Am. Chem. Soc. 83, 4013 (1961); J.A. Schellman, J. Chem. Phys. 44, 55 (1966); A. Moscowitz, Adv. Chem. Phys. 4, 67 (1962); J.A. Schellman and P. Oriel,

J. Chem. Phys. 37, 2114 (1962); G. Wagnière, J. Am. Chem. Soc. 88, 3937 (1966).

[21] I. Tinoco, Adv. Chem. Phys. 4, 113 (1962); J.A. Schellman, Accts. Chem. Res. 1, 144 (1968).

[22] R.E. Geiger and G.H. Wagnière, in "Wave Mechanics, the first fifty years", ed. W.C. Price, S.S. Chissick, T. Ravensdale, Butterworths, London 1973; H. Labhart and G. Wagnière, Helv. Chim. Acta 46, 1314 (1963).

[23] J. Tanaka, F. Ogura, M. Kuritani and M. Nakagawa, Chimia 26, 471 (1972).

[24] A.M.F. Hezemans and M.P. Groenewege, Tetrahedron 29, 1223 (1973).

[25] W. Hug and G. Wagnière, Theoret. Chim. Acta 18, 57 (1970); G. Wagnière, in "Aromaticity, Pseudo-Aromaticity, Anti-Aromaticity", the Jerusalem Symposia on Quantum Chemistry and Biochemistry, III. The Israel Academy of Sciences and Humanities, Jerusalem 1971, p. 127; G. Blauer and G. Wagnière, J. Am. Chem. Soc. 97, 1949 (1975).

[26] W. Moffitt, J. Chem. Phys. 25, 467 (1956).

[27] R.B. Woodward and R. Hoffmann, J. Am. Chem. Soc. 87, 395 (1965); R. Hoffmann and R.B. Woodward, J. Am. Chem. Soc. 87, 2046 (1965); R.B. Woodward and R. Hoffmann, "Die Erhaltung der Orbitalsymmetrie", Verlag Chemie, Weinheim 1970.

[28] K. Fukui, T. Yonezawa and H. Shingu, J. Chem. Phys. 20, 722 (1952); K. Fukui, T. Yonezawa, C. Nagata and H. Shingu, J. Chem. Phys. 22, 1433 (1954); L.J. Oosterhoff, cited in E. Havinga and J.L.M.A. Schlatmann, Tetrahedron 16, 151 (1961); see also ref. [27].

[29] H.C. Longuet-Higgins and E.W. Abrahamson, J. Am. Chem. Soc. 87, 2045 (1965).

[30] N.T. Anh, "Les Règles de Woodward-Hoffmann", Ediscience, Paris 1970.

[31] J.J. Mulder and L.J. Oosterhoff, Chem. Commun. 1970, 305; E.B. Wilson and P.S.C. Wang, Chem. Phys. Letters 15, 400 (1972).

[32] J. Koutecký and R. Zahradník, Collection Czech. Chem. Commun. 25, 811 (1960); T.A. Hoffmann and J. Ladik, Advan. Chem. Phys. 7, 84 (1964); J. Ladik and K. Appel, J. Chem. Phys. 40, 2470 (1964); A. Imamura, J. Chem. Phys. 52, 3168 (1970); K. Morokuma, Chem. Phys. Letters 9, 129 (1971); J.-M. André, G.S. Kapsomenos and G. Leroy, Chem. Phys. Letters 8, 195 (1971); J. Bacon and D.P. Santry, J. Chem. Phys. 56, 2011 (1972).

[33] F. Bloch, Z. Physik 52, 555 (1928).

[34] G. Wagnière and R. Geiger, Helv. Chim. Acta 56, 2706 (1973).

SUBJECT INDEX